SERVO

DAVID GOODWIN

hachette
AUSTRALIA

Published in Australia and New Zealand in 2024
by Hachette Australia
(an imprint of Hachette Australia Pty Limited)
Gadigal Country, Level 17, 207 Kent Street, Sydney, NSW 2000
www.hachette.com.au

Hachette Australia acknowledges and pays our respects to the past, present and
future Traditional Owners and Custodians of Country throughout Australia
and recognises the continuation of cultural, spiritual and educational practices
of Aboriginal and Torres Strait Islander peoples. Our head office is located on
the lands of the Gadigal people of the Eora Nation.

A catalogue record for this
book is available from the
National Library of Australia

ISBN: 978 0 7336 5119 9 (paperback)

Cover and bat dinkus design by Design by Committee
Front cover illustration courtesy of VectorStock
Author photograph courtesy of Natalie Basile
Typeset in 11/17 pt Sabon LT Pro by Bookhouse, Sydney
Printed and bound in Great Britain by Clays Ltd, Elcograf S.p.A.

MIX
Paper | Supporting
responsible forestry
FSC® C104740

To the endless cans of Red Bull, king-sized Cherry Ripes and sausage rolls with too much sauce that got me through those interminable nights.

To every bleary-eyed graveyarder around the world with a halogen tan, pining for the freedom of dawn.

But above all, to Mum.
Thank you for making me love stories. And for helping me believe I had one in me worthy enough to tell.

CONTENTS

How many more nights and weird mornings can this terrible shit go on? How long can the body and the brain tolerate this doom-struck craziness? This grinding of teeth, this pouring of sweat, this pounding of blood in the temples . . . small blue veins gone amok in front of the ears, sixty and seventy hours with no sleep.

—Hunter S. Thompson, *Fear and Loathing in Las Vegas*

I AM THE LIZARD QUEEN

It's eight o'clock on Sunday morning, bucketing down rain. It barrels into the metal canopy, roaring like the vengeful gods of old. Distant thunder growls and far-off heavens sling down lucent veins as fat drops ricochet off the forecourt, spattering those filling their tanks and dragging iridescent stains towards the petrol drain. Inside, the servo is filled with stranded dog-walkers out for their morning paper-milk-and-bread and parents taking their silken-clothed kids to play muddy soccer.

A dishevelled man bursts into the store, shrieking operatic crescendos. He's six feet tall, painfully thin and shoeless, with a straggly red beard and sunken cheeks, clad in a sopping yellow sundress. It hangs off him like a sodden sheet left out on a washing line as he drips dark splotches down the aisles, raving to the music blasting from his headphones.

He's in mirrored, purple-framed aviators, shrieking Kylie Minogue's 'Locomotion' as he struts past the hanging bags of

Minties and Red Frogs, stopping every few seconds to pose and shake his imaginary gold hot pants. Tripping over his own blackened feet, he stumbles into two horrified women and splutters, 'Sorry, ladies, but I'm higher than an eagle's ball sack!'

He skips away toward the deli fridge, as dainty as a ballerina, reverting to Kylie's li'l sis Dannii, but soon tires of her, due to her lack of 'top ten hits and PLASTIC TITS!' He then starts on Delta Goodrem, as he sashays toward the console, oblivious to the old men standing bow-legged, holding rolled-up newspapers as truncheons.

Telling me (through song) that he's lost without me and he doesn't know how he's going to live without me, he approaches the locked glass console door and *face*butts it, knocking his sunnies to the floor. Seen through the wet swords of his mangy red hair, his pupils have dilated to form two gateways to the centre of The Void. He pants on the glass, a canvas on which he then paints – with his purple tongue – a wonky love heart, with an 'I' above and a 'U' below.

One of the dog-walkers approaches slowly. She taps him on the shoulder and tentatively asks him if he could maybe quieten down a bit. You know, because of the kids. He jolts from her touch, hunches over, then goes completely slack. The entire store, from the woman cradling her low-fat milk like a newborn to the slack-mouthed kids grasping their bottles of Powerade with white knuckles, waits breathlessly.

Sensing the occasion, he uncoils, rearing up like some undulant serpent, swaying slowly from side to side then rocking back on his heels before turning to face them and strike, shooting his arms up to the bug-filled fluorescents, rasping, 'I AM THE LIZARD QUEEN!' as he steps on and breaks his sunnies.

As though they were his final Horcrux, he shrieks in world-ending pain, then hops rapidly toward the opening glass doors and out into the downpour, hissing savagely at the petrol pumpers before sprinting off into the rain-soaked morning.

CHAPTER 1

WELCOME TO THE CIRCUS

Hi. My name's Dave and like thousands of Aussies, I've worked in a servo. While this fact in isolation may not appear all that unusual, let alone worthy of an entire book, I'd like to add that I worked late at night. For six years. On weekends. In the far western suburbs of Melbourne. For way too many years to stay anything approaching sane.

Servo work is popular. Everyone seems to know someone who's worked in these halogen parlours of forced smiles and expensive, superfluous crap.

Servos are an unmistakable piece of Aussie life; a part of who we are, like sunburn, backyard cricket and tall-poppy syndrome. Whether it's our friendly refusal to buy two Snickers for five bucks from the comatose student with eyes of sunken craters on some hellish alien moon, an exchange of social niceties as sincere as positive eBay feedback, or simply the comforting buzz as that first can of Monster on Monday morning speeds its dark magic through your synapses, Australians feel at home when they walk in through that parting glass.

A lot of you younger folk may cringe at CCTV footage of yourself early Sunday morning – sloshed and spouting nonsense in the pallid pre-dawn with all the social restraint of Dizzee Rascal – berating some tired soul as you vacuum up stale sausage rolls and chocolate. Meanwhile, your munted, saucer-eyed mate sits in the cab, hugging and stroking the taxi driver who sustains himself only by graphically fantasising about beating the two of you into submission with a Swap'n'Go gas bottle.

A servo is a great place to people-watch. Almost everyone has to fill up at some point, so you get to see all their idiosyncrasies and prejudices in one place. From the sullen, beanied Holden Commodore drug dealers peddling home-grown hydro and dirty speed bombs to the grumpy, take-no-shit 4WD mums who rule their cowed families with an iron fist, everyone passes through those electric doors eventually.

This book is about all those people. From the maniacal freaks and the lilting heroin addicts, to my long-suffering masochistic colleagues, and everyone in between, the following pages detail those many, many long nights and the confoundingly bizarre creatures that made them just that little bit longer.

The book is also about me. Work in a surreal, circus-like environment for long enough and it changes you. It makes you cynical. Unafraid. Manic. Depressed. World-weary. What follows is an addled and somnolent view from behind the counter and under those harsh fluorescents; a six-year voyage of sex, drugs and sausage rolls. Of fake smiles, disabled ninjas, addled wizards and endless drive-offs, and how soaking in this unremitting madness has crafted who I am today. I'll warn you now that it's crazily convoluted and reads like it was written by a confused and irascible penguin with Borderline Personality Disorder. But it's a smashing good time, I promise. Shall we get started?

Getting a job at a servo is not difficult. Pretty much anyone can do it if they're hard up for cash. That was my excuse. It was early 2003 and I'd just started my third year of a mostly worthless arts degree. I was rudderless, and beginning to question the used-car salesman's gleam in my high school career-counsellor's eyes, when he'd urged me to follow my literary dreams; things were not working out as I'd hoped. Poems in obscure literary journals and articles in online magazines paid in nothing but ego and hope, and besides enduring the daily horrors of hours on Melbourne's public transport system to and from uni, I spent my time at home – car-less and penniless – ensconced in headphones and books, trying to block out the disintegration of my mum's second marriage, interwoven with the green fireworks on CNN as Dubya's 'shock and awe' invasion pummelled Baghdad into dust.

I needed to escape. That much I knew for certain. I wanted peace and quiet, yes, but I also craved some crazy adventures, whirlwind romances and awe-soaked revelation in foreign lands that would somehow birth a supernova bestseller to launch my successful writing career. But all those things demanded money and, at twenty years old, I didn't even have my damn licence, let alone a car. I needed to grow up. And fast. All that surrounded me were angry, uncoupling adults with full-time jobs who were loosely funding my partly formed daydreams. I tried to convince my parents that I was an artist, that forcing me to conform to the travails of the real world would crush my creative joie de vivre, but they weren't buying it.

So I skimmed online job ads for something easy, something that paid the bills but didn't require much thinking or hassle, leaving me time to write. Being a child of *The Simpsons*, I became intrigued by a term that kept popping up. 'Console Operator', the ads said, bearing a mysterious resemblance to Homer's role at the nuclear power plant. So I clicked. And called. And showed up to an interview in jeans and an ill-fitting shirt I'd found in the back of my wardrobe.

But even that seemed a little over the top. I was the best-dressed one in the hospital-like waiting room, that is, except for a loquacious, bird-like man in a three-piece suit and shiny black shoes, polished to a mirror sheen. He kept asking everyone's name, how their day was going, and if they wanted a Mintie, as he shovelled them nervously into his mouth, one after the other, his jaw working overtime until he'd created a massive white glob. He kept grinning at us all, with this slimy white ball just behind his teeth, rolling it with his tongue as his gaze danced around the room.

I'd been nervous when I got there, but by the time they called me in, I felt like I could have shown up with a neckbeard and a half-eaten bag of Cheetos and still been successful. I was greeted by a photocopied-looking man behind a chipped desk. He had impressive bags under his eyes. He smiled blankly as he shook my hand and invited me to take a seat. I proceeded to lie through my teeth, inserting such well-worn favourites as

'I possess great initiative', 'I'm a people person', and the sealer: 'I'm passionate about supplying good customer service.'

The successful ones were invited to an information session where, in a tiny room above a city-based servo, we were shown grainy black-and-white videos of stocking-headed speed freaks firing sawn-off shotguns and yelling enraged gibberish at students cowering behind shelves of exploding cigarettes.

The very next day we were carted off to surrounding suburban servos to begin our week's training. My flagrant lie about having previous retail experience was soon exposed, when as I was shown how to open the drawer of the cash register, I didn't even know how to remove the till. 'Another one,' groaned my trainer. 'Let's start from the beginning, shall we?'

They showed us how to count money. Low, out of sight and quickly, like a jumpy bank teller. They showed us around the store, explaining how each machine worked and what we had to do to keep them that way. They showed us which buttons to press, both on the cash register and with people. And how to defuse a situation, like when a burly, tattooed beast burst into the store yelling like a rusted chainsaw that the cost of petrol was 'fuckin' shockin', you criminal pigs!'

More than anything, though, I was amazed at how *fast* the whole thing moved. My years of quiet reveries inside the castles of my own head had me stood back in a bewildered blur as I watched workers hustle about the console like adrenalised smiling soldiers. Stocking smokes, scanning Coke bottles and chocolate bars then bagging them up, processing price changes – their fingers a dextrous blur over the register keyboards – and smiling, always smiling, as the customers rained down hell.

They marched in from the forecourt's river of impatient steel, and before they even got to the counter, they were whingeing about this, arguing about that. Deciding on a product or pump number, paying for it, then changing their mind, only to change it back again and blame the whole thing on the poor console operator. They'd yell questions over their shoulder about their car's specific mechanics from the back of the store while the poor guy or girl behind the counter resembled a multi-limbed deity, serving a line of nine people, restocking the chewing gum, manually typing in numbers from cards that wouldn't swipe and printing phone credit and car wash dockets, all while on the phone to tech support to fix whatever machine had broken down that hour.

I watched, glassy-eyed, unable to fathom how any human could keep their head above water as their brains were assaulted with a barrage of beeping horns, shifting cars, incorrect pump numbers, screaming children denied the treat they'd been coveting, and addicts, in their quiet droning voices, always asking for the toilet key. I figured that maybe once we'd finished our training, we were all quietly palmed a year's supply of Ritalin along with our freshly pressed uniforms and name badges.

I should point out here that working in a servo is kind of like being an all-purpose butler, except that your boss is not a person of wealth or means. Nor is their demeanour particularly benevolent, thanking you for the glass of sherry you bring them as they warm their un-slippered feet by a well-tended fire. Your master, to whom you must declare unfettered smiling subservience, is the teeming populace of your servo's busy suburb. And they will not ask nicely for their supper . . . They will bark their demands as they stride toward you, laughing while you scuttle here and there, trapped in your little anti-jump-wire cage.

Mindful of this unvarnished reality, our trainers only let us serve customers in short bursts, and as we began to fret – mewling like hopeless little kittens as the angry lines snaked around the store – they jumped in and saved us before our heads exploded. Slowly though, we got better. By the end of the week, I actually had *some* rough concept of what I was doing. On the final day of training, I walked in with a swagger and ended my shift with a variance that was less than $100 for the first time.

After I somehow passed my final test, the store manager handed me my freshly minted name badge and an oversized company shirt in plastic wrap, asking me if I'd be confident if I was required to work on my own. 'No worries,' I said, hoping the look I had on my face was something akin to self-assurance, while wondering if they'd really be stupid enough to let a clearly confused and hilariously unskilled me loose on an unsuspecting public.

A few hours later, my retail area manager rang and asked if I was cool with working a shift the following day in an inner-suburban servo.

'Sure,' I said a little nervously, 'no probs.'

'Okay, good. You start at one am. Any problems, give me a call.'

'Wait, you mean in . . . four hours?'

'That's the one.'

Sneaky bastard.

On no sleep, I set off to my first ever graveyard shift. The night was dank with fog, so thick that it poured through the air in a spectral soup. As I walked from the train station toward the servo, the lights of the price board rose up through the murk like some glistering alien monolith.

I walked nervously through the sliding glass – in my giant shirt that swam on my scrawny torso – and was greeted by a well-built man with an impressive mane of hair that shone lustrous in the console downlights. It looked as though he washed it hourly.

Siva was his name – like some proud lion stalking the Serengeti – and he had the exact devil-may-care attitude I was desperately after. He grinned at me and grabbed my hand, giving it a few quick pumps before returning to his customers.

When it got quiet, he gave me a quick tour of the store – which was completely different to the one I'd trained in – telling me to keep the pie warmer full, to automatically safe-drop my $50 notes, what not to touch, who to call if stuff broke down and so on. He must have sensed the desperate vibes emanating from me.

'It's your first ever shift, isn't it?'

'How'd you know?'

'You all have the same look,' he grinned, 'like lambs to the slaughterhouse.'

'Oh.'

Seeing my unease, he placed a friendly paw on my shoulder and gave me a pep talk.

'It'll be fine, mate. Just smile and pretend you actually like everyone. If they're arseholes, well . . .' He placed his other paw on my opposite shoulder, set his jaw then got his game face on.

'Smile again, hold eye contact for *two full seconds*, then tell them to go straight to fucking hell. We don't get paid nearly enough to put up with their bullshit.'

And after adorning me with my duress pendant necklace, he left me with my fear.

For ten or so minutes, the store was deathly quiet; I suppose every hurricane has its eye. I ambled around, strolling through the shining canyons of salt and sugar, squinting under the fluorescents bouncing off their endless colours. Servos are bright, dear reader. Obscenely bright. Villainous scientists have proven that your brain buys much more overpriced detritus of Western civilisation carved from the bones of a dying world if it's fried in an artificial blaze of empty promises so vivid it makes your eyes bleed.

After a while the Starburst, Pringles and Smith's Chips start to scream at you, whirling together into a teeming mass of hyper colour insanity. You often totter, discombobulated, down the aisles, visually drunk on products so unnecessarily bright it borders on belligerent. Golden Crunchies, ruby-red Cherry Ripes and emerald-green Peppermint Crisps glisten on the console's front shelves as priceless jewels atop a dragon's hoard of pilfered treasure.

I snapped out of my daydreaming as I saw my first ever customer approaching. Rushing back to the console to open the locked glass auto-doors for him, I stood up straight, smoothed my massive shirt, smiled my first retail smile and began to greet him. I noticed he was stumbling slightly, nodding and smiling at nothing in particular . . .

My first customer was a heroin addict. Now, when I say 'customer', I'm not being completely honest: he didn't buy anything. He just bobbed there for a while like a narcoleptic scarecrow, making the electronic doors open and close. Open and close. He eventually wobbled back outside and crumpled down by the gas bottles and firewood bundles until the cops came a bit later and picked him up.

More than a little anxious, I re-locked the auto-doors and explored the console. Anxiety began blooming though me as I noticed everything that could possibly be different from my training store *was* different: the buttons on the keyboard, the point-of-sale equipment, the phone credit machine, and almost everything else may as well have been in Klingon. I left the console and headed through the storeroom to the walk-in fridge. After passing through a narrow valley of towering, colourful cardboard, I opened the steel door and was greeted by the cold air and white noise of its huge whirring fans. On my left were hundreds of bottles removed from their boxes and stacked neatly on the metal shelves: Coke, Pepsi, Fanta, Sprite, Red Bull, V, Mount Franklin, Cool Ridge, Spring Valley, Big M, Dare and many others, looking just as bright as the store itself. Above them were messages on several pieces of paper, sticky-taped to the fridge wall and spelled out with red permanent texta, vigorously underlined for sections deemed important:

DO _NOT_ LEAVE HALF-UNPACKED BOXES IN THE FRIDGE!
IF YOU OPEN A BOX, UNPACK IT _FULLY_ OR _NOT AT ALL_!!
NO CARDBOARD IN FRIDGE!

As I looked closer, there were more scrawled messages, none of them subtle:

ONLY _ONE_ FACING OF ORANGE AND MANGO HERE!

And:

NO GLASS IS TO BE STOCKED ABOVE SHOULDER LEVEL!
THE NEXT TEAM MEMBER FOUND DOING THIS WILL RECEIVE
A WRITTEN WARNING!

On my right were the plastic slots the bottles were to be slid into, angled downwards to the front of the fridge. I started grabbing bottles and placing them in the appropriate slots, watching in satisfaction at how a glass bottle would slide down a near-empty slot until it reached the rest with a soft, satisfying *clink*. Along with the humming fridge fans, it was the Tetris-style meditation I needed to calm down a bit, ambling around for several minutes, referencing the angry signs to ensure I didn't get in trouble. Slowly, I began to feel the cold wrap around me and steal its way inside my flimsy shirt. As I stocked the bottles of Pepsi, I realised that you could see snatches of the aisles through the gaps between the drinks. I peered out at the empty store, and something stirred at the corner of my vision.

A large bearded man was standing outside, waving his arms furiously at the door's sensor, as though this extra movement might open it. After glaring angrily into the store, he wedged his fingers into the tiny gap between the glass doors and strained to rip them open. Once this failed, he started shouting, jumping on the spot to see if I'd maybe fallen asleep behind one of the aisles.

Rushing out of the fridge, through the storeroom and back into the store, I looked back to see him glaring at me. I smiled apologetically and ran back to the console to unlock the auto-door. But first I had to unlock the coded glass console door, which had swung automatically shut as I exited, bound for the fridge. Siva had written the code for me on a scrap of paper, which I could see, but couldn't read, as it was face-down in the far corner by the cigarette lighters. Panicked, I assessed my options.

After thirty seconds of standing around feeling like an imbecile and hearing the yelling at the door grow more insistent, I realised that the only way to get back in was to climb over the chocolate bar shelves, the till, and through the taut anti-jump protection wires.

As I got halfway through, wriggling like a fish as the wires strained against me, the irate bleating at the door changed in sound. After my foot caught on the wire and I sent the Eclipse mints promo display crashing to the floor, landing in a crumpled heap in the middle of the console, I realised it was more of a braying. The guy was doubled over in a fit of hysterical laughter. I stood up, unlocked the door and the customer entered, still chuckling to himself.

'Kid, I was gonna give you a fucken serve, but after that display, I don't think I got the heart!'

After much more laughter, he bought his Winnie Blues and departed, still spluttering. As he left, I noticed another person was in the store, having followed the bearded guy in. I looked at him, also with a big grin spread across his face, and realised he was, believe it or not, my Year 10 visual arts teacher. Smiling at me, perhaps safe in the knowledge that his teaching was at least partly responsible for creating such competent members of the workforce, he bought a Spring Valley orange juice and departed with a wry expression.

A little shaken by my own incompetence, I got out the store diary and had a read through the last couple of months. I soon noticed that every second day or so seemed to be filled with anger. There were page-long lashings for not checking the dates of expired milk; for not printing a receipt for a void or a return; for not filling out an incident report form correctly; and, ironically enough, for not reading the diary. The pages were filled with veiled – and not so veiled – threats of unemployment, punctuated by copious underlining in red pen so vicious that I made a silent pledge to try to do everything right; to ensure I'd never be one of those bumbling simpletons who incited such pyrotechnic wrath.

Customers began arriving in a slow trickle and I set upon them, all smiles, aiming to be the best damn servo guy the world had ever seen. Slowly remembering my training, I actually began to get the hang of things, filling my till with coloured notes and genially asking my customers to choose cheque, savings or credit before handing them their printed receipt. Soon the taxis started rolling in and I developed a rhythm, beginning to program my brain with that fake cheery routine that every retail check-out worker comes to know. After an hour of this, I began to breathe properly and feel a little more in control.

It was at about this stage that, after looking at my opened till, I noticed that while I had lots of notes, my supply of $1 coins had dwindled to almost nothing. Retracing my steps from training, I eventually figured out how to get the shadowy-looking change machine to spit out a roll of coins in a tight cylindrical wrapping of cardboard. As I'd observed during training, the experienced operators would smack the cylinder sharply against the counter to loosen its grip, before deftly peeling off the cardboard and tipping the coins into the respective slot in their till. As I was remembering all this, the doorbell rang several times in quick succession, followed by the sharp, insistent tap of keys on glass and I opened the door for an impatient taxi driver with an impressive broom moustache.

While he was counting out his coins to pay for his LPG, I tried the trick of smacking the cardboard hard against the counter, only for the roll of $1 coins to break loose completely and explode into his face. Thinking I was trying to assault him, he started yelling threats in Hindi and grabbed me by the collar, as I offered a startled apology. Most of the coins that didn't end up on the floor were littered across the console counter, mixed in with the taxi driver's, who naturally claimed the majority were his. After several minutes of negotiation, I managed to get something close to what I thought was a 40/60 split, finally getting rid of him.

I was just heading out to pick up the $1 coins when four sets of headlights crawled in off the road, followed by three more, and suddenly a group of drunken middle-aged men burst through the door singing English football chants and then rapidly scoffed all the pies and sausage rolls in the warmer. I walked back to the console to serve them and swore loudly as I noticed that, again, I'd left that damn strip of paper with the code to the locked door beyond sight or reach.

The football fans gave an almighty cheer and accompanying 'NAAH NAAH NAAH!', spilling meat and pastry onto the floor, as I wriggled back through the anti-jump wires like a desperate slug. One of them asked if I was a Scouser trying to rob the place. Once through, I grabbed a permanent texta and scrawled the damn code on the back of my hand as a brand-new chant started up, one specifically concerning my entertaining degree of incompetence.

They finally left but were quickly followed by others, every single one of them eyeing off the gold still lying on the ground as I rushed out in the few seconds between customers, grovelling around on my hands and knees for the remaining $1 coins. I finally grabbed them all, returned to the console and took a deep steadying breath, vowing no more stupid mistakes, before I unlocked the door for yet more customers.

Then I got a blood nose. This wasn't unusual. I got them from time to time, always at the most ridiculously inopportune moments possible, like when I was on stage, speaking my lines during the school play. Or dancing with a cute girl at a nightclub. Or alone, bewildered and afraid behind the console in a bright busy servo. As I felt that horrible trickle start to flow, I laughed darkly to myself, grabbed a tissue and held it there while still trying to serve bemused customers. This worked for about ten seconds, as I quickly realised I needed two hands to accept money and fetch smokes, plus a myriad of other things demanded by my growing clientele. So, stressing big-time now, I grabbed a small wad of tissue and shoved it up the offending nostril with no small amount of malice, as far as it would go. My nose now looked horribly uneven, one side swollen to thrice its normal proportions, the nostril turned up like the snout of a pig, but with a trail of white dangling out of the bulbous mess. Still, I now had two hands back and could concentrate on serving customers.

But before long, the blood soaked through the wad of tissue and the white hanging out of my nose turned red. I started dropping tiny splotches over everything. At one stage, I looked down to see a drop falling straight onto a man's American Express card. It hit the centre and formed a perfect circle, with tiny little rivulets cascading out towards the edges. I glanced up at him, sniffing desperately, waiting for the outburst. But none came. He hadn't noticed. I silently thanked a greater power as I surreptitiously wiped it off and completed the transaction. This happened again and again over the next hour and a half, no matter how much I plugged, wiped, or held my nose. I was starting to seriously struggle.

You see, normally, if you had a blood nose at work, you'd stop working, tilt your head back, and sit in a quiet place with a tissue, and as soon as the blood clots, you're right as rain. But when you're on your own in a busy servo, sitting down unmolested in a quiet place is a three-fold utopian dream. The reality has you ferreting around the console holding your nose, pressing buttons and fetching nicotine for your chortling masters as cars thrum impatiently at the pumps.

By 3 am I was becoming a bit concerned about blood loss. During a merciful lull in traffic, I sat completely still on two milk crates for several minutes, my head leant back against a golden row of Benson & Hedges, and finally the nosebleed began to slow. I then went around the console wiping away the blood splotches while trying not to leave any new ones.

At this point I had only a runny nose that sporadically leaked out a gummy scarlet paste. This was fine, because I could simply wipe it now and again with tissues. But soon enough the tissues ran out, so I replaced them with paper towel, which, to a sore nose, may as well be sandpaper. I checked my reflection in the glass and saw a deranged caricature of Rudolph, wild-eyed and crimson-nosed, with smears of blood drawn across his face in preparation for some deadly Reindeer Games.

Then it got busier. Instead of the odd car rolling in, six would speed in at once like they were entering an F1 pitstop, drivers jumping out of their car and at the pumps, followed by four more, before I could finish gaping at the harrowing velocity of it all. Every time a customer picked up a pump, an ear-piercingly shrill alarm would sound until I touched a square on the screen authorising the pump. With all the cars, the sound was beginning to drive into my brain, but I soldiered on, stabbing my index finger at it again and again without even looking at the screen.

As the latest rush departed, I noticed two remaining amounts of fuel still blinking on the screen, but with no cars at the respective pumps. Thinking they'd moved away from the bowser to use the toilet, I waited, but soon realised I had my first two drive-offs. This wasn't good. We'd had it drilled into us mercilessly during training that we were responsible for what happens on our forecourt, and drive-offs – especially with no recorded numberplate – were bad news for our employment prospects. Starting to feel more and more uneasy, as the conveyer belt of customers continued rolling in, I fell back on my strengths – garnering pity and looking confused. I started panicking and losing stuff: pens, paper, patience.

Then the bank lines went down.

Anyone who's worked on a register in retail will know that the only customers that seem to carry cash are taxi drivers, drug dealers and those old men who like to jingle small denominations of it in their pockets with maddening grins on their vacant faces. People pay by card, and so did all my waiting customers.

As every customer approached the console and pulled out their card, I apologetically directed them to the ATM and they directed their anger and complaints right back. 'That's the third time this week, this place is a *fucken joke*!' They threw their cash on the counter and spitefully refused to tell me their pump number, soon leaving me with a dirty pile of cash and a screen full of flashing numbers. Not knowing who was who, I let it pile up until I had this little mountain of money. It would've been paint-by-numbers for anyone who burst in with larceny on their mind.

Then the ATM died, haemorrhaging out its final payload to a clubber before blue-screening. I stared at its blank 'OUT OF ORDER' message, mocking me from afar, and bright panic bloomed in my guts as I realised I was in for a final few hours from hell.

I pushed it down and fake-smiled at the next approaching customer, who pointed at the empty pie warmer with pure malice in his eyes. After rushing to fill it, I returned to the console and, reaching into some dark corner of my memories from training, I found the manual swiper for cards. It looked like some rusted Soviet artefact of war. By this stage it was 5 am and the store was chock-full of early-risers marching in impatiently off the forecourt, opening then slamming the fridge doors as they filled their arms with sugar, newsprint and caffeine, the console shrieking every few seconds as another hand grabbed a petrol pump, the forecourt a shunting mass of coloured steel, but I soldiered on, working up a nervous sweat swivelling round every thirty seconds to grab smokes, lighters and printed pre-paid mobile credits in between manual swiping as the line of impatient customers grew ever longer. (I found out later that I'd put almost every manual transaction on the wrong form and subsequently lost the company over $6000.)

I got a call through for a fuel price change at 5.30 am and sprinted out to manually change the price board. As I returned to the store I met the furious whites of dozens of smouldering eyes in a long, long line of cross-armed people waiting for me inside, the queue already snaking its way out the door.

I kept mumbling feeble apologies for my ineptitude ('Sorry, it's my first day'), but it didn't seem to help. Customers' sighs, mutterings and obscenities grew louder and more belligerent as I fumbled incompetently with the swiper, my pen a nervous blur scribbling litres, dollars and odometer readings into boxes, as a small creek of sweat snaked coldly down my ribs. The swiper kept jamming so, to get a proper imprint of the card you really had to give it the beans, and when you added people's slightly bent cards into the mix, I managed to irreversibly mangle six of them, cringing while handing back plastic origami interpretations of their main means of payment along with furiously requested customer complaint forms. I started to properly unravel.

You can't quite communicate how constant and unrelenting a new, poorly trained solo customer service job can be to someone unless they've actually worked in one. Your anxiety becomes this sharp glinting beast, clawing its way up through you as your mind begins to fog over in terrified confusion.

And servos are infinitely worse than any Macca's, Coles or Kmart. At those places, you can at least defer your problems to someone nearby. At night in a servo, you're always on your own. Out of options before you've even begun, locked in a little cage and bathed in searing halogen. You can't 'call the manager', because she's fast asleep, like normal people are at 4 am, so you have to face the monster down – again and again – until you're finally done.

And the monsters don't stop coming. Like angry oversized lemmings, they trudge on in, one after the other. They stare at you piteously while they wait, too, with all the carefree malevolence of Blofeld stroking his cat, so you can't escape the groans, the rudeness, the impatience, the piercings, the six packets of Tim Tams, four bottles of 2-litre Coke and three tubes of lubricant cradled in their arms, even while you're militantly monitoring the actions of every motorist on the forecourt, accepting fuel deliveries and bagging up all the endless overpriced garbage you'll soon have to restock if they ever go away. Three years after I started, I still struggled with it. For a first-timer it was hell.

When the store manager finally came in at 7.45 am I was overjoyed. Squeezing past several folds of the gargantuan line bent around the store and seeing my wild hair, bloody war paint, and the console looking like the bombing of Dresden, she took a motherly kind of pity on me and immediately reset the system, hushed the merciless beeping, reconnected the bank lines, and started rapidly dispersing the Great Wall of China–size queue of incensed customers now late for work, even though she usually worked only in the office.

My shift officially finished at 8 am, but after finally collating my endless blood-smeared receipts, marvelling fearfully over my shift's negative $459 variance, and filling out six drive-off reports, I stumbled out the sliding glass doors at 9.20 am with fear-stained armpits and tottered to the nearby train station in some adrenergic haze, wondering if this was what the world of employment was really like.

Two more nights of this and I was like a skittish cat hooked up to an IV of some experimental stimulant. My dreams were Dali-esque juddering nightmares of elongated taxi drivers, singing meat pies and cash registers cackling like hyenas, black tongues slurping from their darkened maws full of jangling metal teeth. Several times I woke in the afternoon in a cold neon sweat, fearing what new catastrophes future nights would spew at me.

As I was drifting off to sleep on the train home from my third shift, my retail area manager rang. Expecting to be fired, I woke up instantly. Instead of sacking me, though, he offered me permanent shifts in my home suburb, four a week, but only more graveyards. It should have been an obvious clue to me that this organisation was (1) horrifically understaffed, and (2) wasn't exactly picky when it came to replacing them, but, amazed I still had a job, and knowing how busy servos were during the day, I accepted instantly, hung up, then fell into an exhausted sleep. I jolted awake an hour later at Flinders Street Station, having missed my stop, with the train heading back the way I'd come.

After I eventually got home, it dawned on me that I had no idea how I was going to squeeze in work and full-time uni with next to no sleep in the middle. I'd always been a crazy night owl, but this was in all reckoning probably a bit much. But, as happens so often when ensconced in the impenetrable armour of youth, I promised myself I'd somehow make it all work. To put it out of my mind, I blearily started my essay on the Chinese Revolution of 1949, due that very same day. I collapsed into bed within fifteen minutes and slept through class. It was a sign of things to come.

CHAPTER 2

DAVID AND GOLIATH

Because I figured you'd ask, dear reader, I lived in Werribee.

You may have heard of it. Like many lower socio-economic areas, its reputation undoubtedly still precedes it. Googling it at the time of writing, even though it is nowadays more a burgeoning shiny metropolis than a battlefield of warring teenage gangs, the most popular searches are 'Is Werribee dodgy?', 'Is Werribee safe?' and an enduring favourite, 'Does Werribee smell?'

The last of these is due to the existence of the Werribee Sewerage Treatment Plant – or as the locals have long termed it, the 'shit farm'. Over a century old, it recycles half of Melbourne's porcelain offerings annually, wafting, some apocryphally say, a pungent bouquet inland from the coast.

I'd lived in Werribee and its surrounding suburbs most of my life, and its streets, the same ones Google strongly recommends not to walk alone at night, had provided an education markedly different from the one I'd received at school. Now that I was facing long nights exposed to its nocturnal population, I couldn't help but wonder what it had in store for me.

A clue may have laid in my town's original meaning, the Wathawurrung name for 'backbone'.

God knows I needed one, as my mum dropped me – twenty years old – off at work just before midnight with a kiss, a packed lunch and promises to be extra careful.

Walking across that darkened carpark toward the servo I'd known so well – having stopped off after high school every day for years wasting what scant money I had on piles of bright sugary crap – was an eerie feeling.

As I walked through the juddering auto-doors, I was greeted by a lanky young man in a well-tucked shirt with a friendly disposition. 'Ralph', as his name badge said, was an agreeable fellow, engaging every entering customer in meaningless but pleasant chit-chat as he efficiently processed and bagged up their purchases.

He unlocked the console door and welcomed me in and we chatted between midnight customers. Though only in his mid-twenties, he had a kind of doddering, grandfatherly air about him. He was extremely helpful as he showed me around, giving me a thorough tour of what was to be my home store, which, thankfully, was similar to the one I'd trained in. He spent ages on every machine, explaining their functions and then getting me to repeat them, to make sure I'd been listening and fully understood their operations.

Before I knew it, it was 1 am. Ralph, having graciously stayed an hour later than he'd needed to, unpaid, finally left, but only after he was satisfied that I was cool with everything.

'People are pretty normal around here,' he said as he walked out the doors. 'You shouldn't have too much to worry about.'

Knowing Werribee, I was cynical at best, so I hung my duress pendant snugly around my neck, locked the doors, and prepared myself. It was quiet for a while, but as the night dragged on, people started showing up. Some customers *were* normal, buying a pie, a Mars Bar, or a bottle of Coke then quickly leaving, while others would speed past the pumps, whooping like madmen as they fishtailed across the forecourt, making me duck to the floor, waiting for them to smash straight through the gas bottles and immolate me in a dirty bloom of orange flame. I'd look up and they'd be stopped inches away from me, smiling.

Heaps of kids came in trying to buy smokes, giving me all sorts of fake IDs. One couple came in and bought a DIY pregnancy test kit. The desperate guy and his girl did the test in the outside toilet, running back in several minutes later to tell me the news: the oven had a bun. Seeing the looks on their faces, I wasn't sure whether to congratulate or commiserate with them. An aquamarine VW Beetle full of people on a wild goose chase for an ice cream Crunchie was next. I was their seventh store. After them came a gold BMW full of unnervingly happy people pumping Roxette's 'Joyride' so loud it filled the forecourt like aural rain. One of them, wearing a balaclava, stood up through the sunroof and started launching water bombs at unsuspecting customers that turned out to be filled with different flavours of cordial.

They kept coming. In clumps. One thing I came to learn was that customers, like the far-off planets they surely originate from, have their own gravitational pull. While you may get no one for an hour, if just one car rolls in, they are promptly followed by another five. This will happen, without fail, night after continuous night. It is one of the inviolable natural laws of the Servoverse.

Next was a guy who came in saying, 'I want to get rid of a bit of change if that's okay?' He disappeared into his car and came back a few minutes later with an old brown 70s 'stable table' (the ones with the beanbag-like base) covered completely in what turned out to be $294.75 of mostly silver coins. He then drove away in a more fuel-efficient car, while I was stuck with enough metal to melt down and make into a tank. My till was so heavy it took two full seconds to open.

I slowly got the hang of things, introducing myself to the local taxi drivers, who grunted and demanded a receipt and the toilet key. I asked one of them, an old Frenchman who returned from a noisy twenty-five minutes, whether the end result was a boy or a girl. 'Twins,' he shot back, before chucking the key at me and skipping back to his cab, several kilos lighter.

I ambled around the store, checking out all the obscenely bright, overpriced sugar, chocolate and salt. As it approached 2 am, it seemed like it was beginning to quieten down, so I got out my history textbook and started on my essay, which was already a day late. I'd gotten halfway through my rushed introduction of Mao Zedong's heroics when a massive truck pulled up and a gruff, hairy man in a fluoro yellow singlet dragged in a pallet piled high with endless boxes of Coke, Mount Franklin, Fanta, Snickers and Twisties. As he set it down next to the coffee machine, I gaped at all the stock they expected me put away. The essay would have to wait. He smiled evilly as he glided toward the door on his pallet trolley, like a kid on his new Christmas scooter. The smile remained on his face as he returned with another two pallets, each piled higher than the last, along with several large boxes of cigarettes, which he threw at me like a rugby ball, laughing to himself as I tried not to stumble back into the cigarette cabinet with each new volley of tobacco.

'A new kid, eh? Better get some meat on those bones if you're gonna be doin' this every weekend!'

As he glided away with the same spiteful grin, I stared hopelessly at the city of cardboard towering over me. *They seriously expected all this put away on a Friday night?* Not wanting to be defeated, I attacked it, carting box after box to the store's empty waiting shelves, remembering the FIFO – First-In-First-Out – method of stock rotation they'd imprinted on our minds in training. I took the boxes of drinks into the fridge, ripping them open, placing the bottles on the metal shelves and in the plastic slots, before flattening each box and adding them to a pile growing larger by the minute.

The problem was, though, every time I got halfway toward the storeroom, a clubber, drunk or general misanthrope would show up, making me drop the box and scuttle back to the console, ready to serve. If you've ever shown up at a servo at night and were greeted with a locked door, a two-minute wait and a grumpy servo worker, this is why. Either that or they're blearily trying to explain why Chinese Nationalist troops were so quickly crushed by the People's Liberation Army. Probably something to do with sleep deprivation.

These interruptions went on for two hours, exacerbated by deliveries of milk, bread, newspapers and magazines, along with countless taxis, mangy kids on BMXs, security guards, dead-eyed weirdos and other people out at four o'clock on a Saturday morning. So many of them couldn't get over one little thing: I was white. Every second customer would bellow (jokingly): 'My God, an *Aussie*, what time does Habib start?' I would answer (truthfully): 'In about four hours.'

But before Habib showed up, more adventures were to come. By about 4 am, I'd put away most of the deliveries, and was in a semi-daze, walking around trying to remember the seventy-four different things Ralph had studiously shown me. One of the things he told me, several times, was to make sure the door was always locked.

As the thought occurred to me, the auto-doors juddered open and a man mountain burst in, shouting several colourful obscenities. He was as big as Goliath, an absolute hulking monster, in every sense of the word. His tattooed arms bulged from his black singlet, his wet black hair was plastered to his simian brow, and blood was trickling down his neck from a nasty-looking cut.

'Who says crime doesn't fucking pay, *HEY BUDDY*!?' he growled at me, his eyes glinting like nuclear reactors. Reaching a bloodied fist into a pocket of his shorts, he showed me his spoils: a few $50 notes and some cheap-looking jewellery.

'Fuckin' *PIGS!* Won't catch me this time,' he snorted. '*No sir FUCKEN REE!*' He lit a cigarette and studied me closely as I tried not to make eye contact or breathe. 'You didn't see nothing, champ. *Got it?*'

I'm not sure if he realised the seven different mic'd-up cameras on him saw and heard everything, but I decided he didn't really need me to tell him that.

He stalked to the deli fridge and grabbed three roast chicken rolls, which he threw into the microwave and put on high, meanwhile finishing his cigarette in record time and crushing the filter into the grimy tiles. He then walked to the fridge and drained a Chocolate Big M, then an Iced Coffee, one after the other.

Next stop was the porn section, where he grabbed a copy of *People* magazine and brought it to the counter, leafing through it and commenting on the girls as though we were old mates catching up over a beer. After turning to the 'Home Girls' section, he pointed to a girl named Shyla who was from Wagga.

'Fucked a girl from Wagga once. Well, more than once really. Well, she fucked me. And everyone I knew. Found out about that later though. She was in *People* once. Well, twice actually. Last I heard she moved back up there. Wagga. She had a brother. This skinny guy. Drove go-karts. Weird looking, superhero-costume-wearing-muthafucka. Crazy hair but knew every Captain Planet episode off by heart.'

Goliath seemed to be on some kind of drug, as he paced the front of the console and blabbered away almost incoherently. He gave me his views on the government (he thought distant ex-premier Jeff Kennett was still in power, 'and doin' a fair job too!'). On the subject of society, he thought all drugs should be legalised. His opinion on the youth of today: 'Wasting their lives away with drugs and alcohol and Jeff had better damned-well do somethin' about it, the fucken tiprat!'

Then he started on me. 'You look like a good kid. You ever done any drugs? I have. Hard drugs. Fast drugs. I reckon I've had more drugs than you've had hot dinners, champ!' He pulled out a tiny bag of white powder and tossed it on the counter.

'I can pay with that, if ya want. Just have a line though, don't have it all. Ya might blow the place up!'

After I politely refused, he tried to pay with what looked like an old woman's brooch: 'Ya sure? There's real diamonds in there. I think. At least they look like diamonds.' He held it up to the fluorescent lights and squinted. 'Maybe they're pearls . . . Nah, diamonds, definitely diamonds.'

'Sorry mate, I can't accept it,' I said.

He glared at me for a second, flaring his great nostrils like a bull about to charge, then eyed my duress pendant, and quickly brightened.

'Aye aye, Captain Birdseye! Cold hard cash it is.'

Right then, a mop-haired kid rode up on a bike, getting off and leaving it next to the auto-door. He ambled into the store, laid eyes on the beast and froze. Goliath shoved the brooch in his pocket and grinned at the kid, all teeth.

'G'day champ, whaddya doin' out this late? If I was your mum I'd fucken skin ya.'

The kid opened his mouth to talk and eventually came out with: 'Slurpee. I came to get a slurpee.'

'On a freezin' night like this?' Goliath screwed up his face. 'Kids today, fucken loons they are.' He turned to the kid. 'Go on then, get ya fucken slurpee.'

The kid raced to the slurpee machine and filled his cup, and I could see him willing the frozen goo to pour out quicker. He rushed to the console, dropped a $2 coin down, then made a break for freedom, but was called back by the monster.

'Hey kid, gimme a taste of that slurpee.'

The kid statued, before turning around and slowly walking back to his tormenter, eyes on the ground. Goliath grabbed the slurpee, picked up the spoon straw and daintily dropped some into his mouth.

'That's the stuff kiddo. Now piss off and go home. And stay in school. And don't do fucken drugs alright. No fucken drugs. They'll murda ya.'

The kid shook his head vigorously and half-ran to his bike, before pedalling off madly into the dark.

Goliath watched him leave, then jerked suddenly, ripped from his reverie by a roast chicken roll in its unopened bag exploding in the microwave. As though this was the impetus he needed, he grabbed a king-sized Mars Bar and pointed it at me.

'Have you ever pissed in a chick while you're fucking her?' he asked me, as he ripped it open and hoed into it. 'It takes a lot of fucken concentration, let me tell ya.'

CHAPTER 3

MIDNIGHT MADNESS

Somewhere in the general vicinity of my servo exists a tear in the very fabric of space-time. A portal has opened between our world and another, one untouched by light, civilisation and deodorant products. From this void emerge wretched, dishevelled creatures bearing little if any resemblance to you or me. Attracted by the luminous glow, they approach slowly, long faces swimming out of the darkness. Their eyes grow wide with wonder as they observe the glass parting for them, as if by some beguiling sorcery. Once inside, they shield their gaze from the searing fluorescents and shouting colours, palms pressed fiercely against their temples as they try to process this new rush of sensory input.

Then they ask me for a 'packeta Winnie Blues fanx champ' and rejoin the murk from which they came.

After too many years of working in them, I've concluded that servos will always attract those a few litres short of a full tank of petrol. Seriously, in what alternate universe would a Commodore fishtail in through your *exit* – tyres yelping like an injured dog as it backslides toward the glass and waiting gas bottles – only to stop just in time for a bald guy to burst from the passenger door to ask desperately if you sell any desiccated coconut?

As I continued with my first few weeks of the job, I came to realise that a lot of this weird stuff seemed to happen around midnight, just as I was starting work. I began to develop a theory for it: in both European folklore and African voodoo, the hour of midnight is a significant one. It's known as the time when certain gates between this world and another are unlocked. Supernatural creatures are thought to be at their most powerful, and black magic at its most effective.

As you arrive a couple of minutes before midnight, the store that greets you is almost eerie, camped in silence as you walk from your car toward the bright oasis of lights. After some friendly chit-chat with your departing co-worker, you gaze calmly at the empty forecourt, looking forward to a quiet night, but just as the green figures on your screen turn to 00:00, an army of lights, like the lambent eyes of creeping demons, begins to roll toward you. Soon the forecourt is full, seething with a dark circus of broken souls and their varied means of conveyance hurling autotuned dissonance out into the night. They will then pack the store like some impromptu Halloween party for the next half-hour, decimating your neatly stacked aisles and jabbering like a troop of malevolent baboons. Only after causing maximum chaos will they then disperse, vanishing as abruptly as they arrived, leaving you staring, once again, at an empty forecourt. I call it the Cinderella Theory.

While it often bunches up at midnight, you've probably already gathered that weird stuff happens all night if you're a graveyarder. It was about 3 am, a few weeks after I'd started and, having just finished a stale but free bacon and cheese pie and the previous day's footy section of the *Herald Sun*, I was filling out some shift paperwork at the console and happened to glance up at the forecourt.

A dead-eyed man approached slowly, materialising from the depths of the inky night. In his arms he held a loaf of bread, rocking it from side to side as if it were a precious infant. He ripped chunks from it absent-mindedly, letting them fall to the ground in a Hansel-and-Gretel-style trail. He wore a pair of tight black jeans, which wasn't a great look for his emaciated legs. But it was his face that got me. It managed to look both vacant and pissed off at the same time. He entered the store and stared at me with a deadness that felt contagious.

Rake-thin, with sunken cheeks, he walked past me to the chips aisle, ruffling packets for his own amusement. After about a minute, he approached the register, pinned me down with his black, iris-free eyes and asked me, his voice surprisingly clear, 'Do you know where madness is?' Without waiting for an answer, he told me: 'Wherever you find it . . . Or wherever it finds you.' And with that he walked out into the dark.

Later that morning, as I went to stock the chips, I saw that every packet on the upper shelf of the aisle had been slit open.

Freaks. You never get used to them, no matter how many years you spend marinating in their unbridled madness. While some can be amusing, others make you feel marooned on a desolate wasteland of a planet in the far reaches of a toxic crimson Oort cloud, under siege by a collection of abominable life forms, each more repellent than the last. Sometimes their markings are obscure, revealing only snatches of their waiting delirium. Like a bald, spindly man on a bright green Kawasaki Ninja, who strode into the store in a cream-coloured robe with nothing on underneath, and after chatting with me for twenty-five minutes, told me I was ripe for Satan's mansion of fire, leaving a 'Do You Know Jesus?' brochure behind on the counter.

About six weeks into the job, and in keeping with the freak theme, I was having late-afternoon beers with my mate Stevo in lieu of much-needed sleep. Well, *I* was having beers. Stevo favoured the broader brushstroke of a 1-litre bottle of cheap vodka and Fanta, poured into a massive stein he used for such endeavours, in an eye-watering 40/60 split. We were in my backyard, underneath the gazebo. The autumn sun had slid like a cracked yolk down the heavens, coating our faces orange and fierce, lemon scraps of cloud splayed across the sky like the lucent innards of a leviathan.

Stevo is . . . Hungarian. I mention this only because Hungarians, I'm told – and indeed, my experiences seem to confirm this – are somewhat insane. While this madness can take many forms, their utter divergence from normal human beings is perhaps best summarised by Nobel Prize–winning Italian physicist Enrico Fermi, who, when asked by his acolytes in California if extra-terrestrials exist, replied, 'Of course – they are already here among us. They are called Hungarians.'

Stevo is huge, hairy and covered in many large tattoos. He attacks everything in life with a jumble of extraordinary resourcefulness and naive unknowing, and it is often difficult to tell which. A zany, outwardly menacing, but soft and empathetic plasterer with a highly developed skillset of smuggling alcohol about his bodily person, he takes pride in free-balling, renouncing underwear as restrictive and unnecessary. He considers the toilet – one of his consuming passions – his 'office' as, due to some rather abnormal gut chemistry, he spends an inordinate amount of time seated on one every day, leaving offerings to challenge even the most powerful plumbing systems. Most of his phone calls to me were from the Office, while he was busy 'snapping off a grogan'.

Such a term belongs to a language he has developed, a mutant mix between English and Stevo that he's christened 'Stinglish'. I've spent a decade becoming proficient at Stinglish and am given new words every time I see him. During our friendship I have added Stinglish as a language to my word processor. It is currently 278 words large and growing.

Stevo drives a searingly bright, ancient orange ute with a megaphone-style horn that sprays the surrounding suburbs with programmed farm animal noises. When excited he drags in air with great whooshing gasps and you can almost feel the trees bend. He will then lift the person nearest to him and hold them high above his shoulders like a trophy, crowing before concerned onlookers, like an oversized rooster hepped up on goofballs.

Stevo's laugh is more infectious than Omicron. Equal parts hyena, crazed mountain gorilla and drunken witch who's just won the Powerball jackpot, it's not just a cackle, it's an upsurging of chaotic joy that begins from deep in his stomach, bubbling up within him by way of some unseen force, bursting into a sound that conquers all, as he explodes in jubilation with every single one of his alien atoms. It sends him crumpling to the ground like a Jenga tower in paroxysms of stomach-shuddering convulsions. Or it tips his head back to roar like an orchestra of mutant kookaburras at the sky. It is the best laugh in the world, and every time I hear it all I want to do is find a way to make it go on forever.

I was recounting tales from my first few weeks on the dark side of the moon and after hearing a few of them, Stevo leaned back in his chair, deep in thought.

'Sounds like you're surrounded by Gumbletons.'

'What the hell's a Gumbleton?'

'Did you watch the cartoon Gumby when you were a kid?'

'Umm, yeah, I've seen it.'

'What did you think of him?'

'Gumby?'

'Yeah.'

'I dunno. He's just a . . . actually, I don't know *what* he is.'

'Exactly. I've never trusted him.'

'*Why?*'

''Cos he's a Gumbleton.'

'. . . I see.'

Stevo took a big swig of his Fanta Firewater.

'A Gumbleton is someone with no hope in life. A white crayon; a chocolate teapot; an ashtray on a motorbike. They're not good for anything but equally you never know what the fuck they're gonna do. And you, my friend, are surrounded by a herd of wild Gumbletons, which makes you the Minister for Gumbletania.'

And with that, he raised his orange stein to the last of the sun, crowing like a rooster at the sky.

'Seriously though, I'd get out of there while you still can, before you turn into one permanently.'

While I was a little scarred from the oddities I'd already suffered, I scoffed and donned my by-now-familiar air of confident self-assurance and said it was the easiest money ever: no boss looking over your shoulder, free food everywhere, and getting paid to sit around and read soft porn. 'People are people,' I said.

Soon enough, though, I learned to understand insanity.

As the shifts began to pile up, I spent the small hours of the night trying to catch up on overdue uni assignments, but again and again I found myself stuck in my little anti-jump-wire cage, frozen like a jungle rat as several king cobras wove toward me, hypnotising with their strangeness. There was this woman who came in every Sunday night, usually around 3.30 am, always buying a different chocolate bar. Never the same one twice. She always completely ignored me – never making any eye contact, and always in a deep paranoiac discussion with different versions of herself. There were at least three distinct personalities in there. I called them 'Good Cop', 'Bad Cop' and 'Strange Cop'. For the first few weeks I was so flummoxed I just stood there silently. Soon enough, I tried to engage her with pleasant chit-chat, but this only seemed to make her own internal dispute louder and more vehement.

It usually went something like:

'G'day, how are you?'

'. . . they never understand, not when it really comes down to it. I try to tell them but it whispers, just whispers in the dark, and then they leave. No, no, they don't understand!'

'No! I of all people understand, you must believe th—'

'You lie, but not well enough to fool me. No, never to fool me. There are few who understand. Wisdom is rarer than you think. Why would one consider such deception, unless . . . ?'

She would go on like this for several minutes, with her tightly bunned head down, less than a metre away from me, scanning the console shelves' glistening jewels of commercial cocoa for a brand-new offering, eventually placing a Cherry Ripe, a Turkish Delight or a Twix on the counter, always with a $5 note.

Strangeness arrived in various forms. A few Sunday nights later, I was putting some pies in the warmer ready for the morning rush when a plate-less VN Commodore drove up over the front walkway outside the store, so that it was centimetres away from the locked glass doors. The driver, totally expression-less, began to nudge them with the nose of his car, accelerating ever so softly, bending the doors inward as they began to creak under the pressure. His face didn't change. He didn't even look at me. He just stared straight into the store before reversing and driving off.

I learned to map their approach as, struggling with an essay on international politics or social studies, I'd hear the dying whine of a motor, and emerge from the back room to see a bald loon pull up in a struggling green Datsun 120Y, trailing behind him a cloud of smoke and sounding like a lawnmower with a tennis ball stuck in its blades. The psycho would exit his car and, once inside the store, start stalking the aisles, gulping down a can of Coke from the fridge and munching on a pine-apple donut while trying to sing 'Bohemian Rhapsody', burping half of it up and spreading fizzy donut juice over everything.

Unlike crazed bogans, the local heroin addicts weren't particularly animated. I thought working in the outer suburbs would mean I'd avoid them, but smack was seemingly ubiquitous. Most of them only mumbled and drooled, stretching everything out into the slowest malaise. Zombies on Xanax. One of them looked a hell of a lot like Beavis: a scrawny frame with pencil-arms, a long head and a pointed nose. His boof of blond hair sat atop his head like a pile of used carpet. Looks aside though, he was as far away from Beavis's fist-clenching hyperactivity as you could get. He'd droop downward as he walked into the store, just after shooting up. He'd browse the aisles at a glacial pace, grinning at the rainbow-coloured junk food as the junk itself sang golden through his veins.

Twenty minutes later, he'd sway toward me with a family-sized block of Cadbury's Snack chocolate, slowly lifting his head to smile, before placing the block on the counter and dropping a $20 note on top, as his head nodded back down again, deep into his cloying velvet dream.

He'd then – slowly, like a sloth on a benzo drip – tear it open, one tiny rip at a time until, after several minutes, the chocolate was finally freed. I'd softly pick it up by its edges and place it over on the counter for the second till so he could enjoy it unmolested, and I could continue to serve bemused customers.

But every now and then, things would switch up. A temporary heroin shortage meant beggars couldn't be choosers, and so ice took its place.

One morning, as the forecourt traffic was starting to build, an enlivened Beavis powered in on a girl's pink BMX, now much closer to his alter-ego's maniacal vibe. Like a hyperactive shark, he began to encircle the early morning customers pumping their petrol. He closed in on his prey, bobbing his carpeted head up and down like some wired chicken. Closer and closer, ringing the bike's bell and whooping like a madman – until he rode into the back of a parked ute and sailed half into its rear tray: a crumpled mess as his chicken legs hung over the side. The driver came back from the store and started his ute. Oblivious to his new passenger, he stopped only because of the shouts of others on the forecourt.

CHAPTER 4

A REQUIEM FOR GOON

As you can imagine, this sort of thing was pretty full-on for a cloistered little mummy's boy only a month or two in, who'd never really had a job before. While I dealt with several freaks on the train to uni every day, they were mostly tame, and I could change carriages if I wished, which I often did. But here, stuck in my cage, it was almost as if the fluorescent lights of the servo charged them up somehow, transforming them into some superhuman breed of freak. Soon my frazzled nerves started to reach breaking point. Desperate, I asked Ralph one night as I came in for work what he did when he encountered 'difficult' customers.

'Oh, you mean, like people who want one chocolate bar, then pay for it, then they *change their mind*? That really gets on my nerves.'

'Uhh, not really. I suppose I mean more like weird people who won't leave the store when you need to get some work done.'

'Oh, those people. I usually just approach them and ask if I can help them with anything. That usually sorts it right out.'

The following night I put Ralph's advice into action. A short man with a weathered face had entered the store over twenty minutes ago and I had read the same paragraph of my short story for class seven times while waiting for him to leave. I left the console cage and approached him gingerly, asking politely if there was anything I could assist him with.

He jumped, turning to face me with fear wrought across his wizened face that quickly morphed into a glazed but fierce stare.

'The holes are there, they're just hard to see. Can you see them? Sometimes they move, sneak up behind you and try to suck the air you were just breathing. They're sneaky like that.'

I stared into his craggy face and noticed that, while he was looking directly at me, he also seemed to be looking at some point beyond me in the far distance. Little did I know, I had just encountered the biggest freak of all.

Randall was a shrivelled, gypsy-looking man in love with a sun that left no shadow. Forty-five, but looking more like he was sixty, his tobacco-stained hands were marbled with dark splotches, so the ends of his fingers actually looked leprous, ready to drop to the floor any second. He was hopelessly addicted to cigarettes, which made him an addict too, I suppose, only his drug was legal and could take decades to kill him.

His face often seemed perversely cruel – twisted into an expression of perpetual cynicism – so that even when he smiled, he managed to look like a man who had seen the slaughter of thousands on bloodstained fields . . . and sometimes enjoyed it. Google Nick Nolte's Hawaiian-shirted mug shot when he was sideways on GHB and God knows what else. It's pretty much a perfect match.

Randall would spend hours talking nonsense at me, peering into a far-off galaxy, always searching for his black holes. His ramblings were of wizards and auras and his supposed journeys to far corners of this globe and others. These stories never ran out. Every few nights he returned with more, his rantings peppered with enough lucidity that, after a few months, I was almost convinced he was some kind of shaman.

Randall had, in fact, found God. Or some loose approximation. While he called himself a 'mystic', he attended mass at the local church several times a week, sparing what little he could for the collection plate. Sometimes he would march into the store, eyes shining with religious fervour as he shouted passages from the Bible melded with his own delusions. Here his speech was eloquent, laced with words only a well-read man would even attempt to use. Other times, he'd skulk in flat as a tack, his dark eyes windows to the blackness sloshing within him.

On these nights he seemed fully aware of his situation; awake to the hopelessness of his never-ending routine of smokes, church and filling the lonely hours of the night. He'd stare at the shiny cars and young offerings of flesh with a furtive longing, trying to bum smokes off those with smiles on their faces. When he scored one, he'd light it instantly, drawing the smoke into his lungs with a shaking power.

At least once a week, he'd come in and try to buy smokes with a seemingly never-ending supply of DVDs. Now and again, I'd accept some and buy him a packet of 30s. Or I'd buy him a packet on credit, which he would unfailingly pay back the following week.

Randall would sometimes carry a cherry-red acoustic guitar. He'd walk in at 2 am and take up residence by the second till, strumming and proselytising his unique brand of cosmic fire and rambling brimstone at every entering customer for the next two hours. To break the titanically weird ice, I'd ask him to play me a song I might know, but all I ever got were his own creations. They weren't whole songs, exactly; more fragments that were jarring – but also somehow beautiful arrangements that often stayed with me, long after my shifts ended. Once, on one of his more lucid nights, I asked him where his music came from and he stopped strumming, considered for a second, then told me he'd open each chamber of his heart to Jesus and then his fingers would begin to dance with the spirits, like tendrils of smoke over a virgin fire. 'I usually play with a light blue aura,' he said. 'It's better that way.'

He'd cut a lonely figure on the forecourt at 3 am on a Saturday night, walking toward customers filling their cars, strumming the guitar with a beatific look of hope on his ruined face. Usually dressed in a ratty black trench coat or beaten leather jacket, he resembled a shrunken, decrepit Johnny Cash; a suburban, delusional Man in Black.

Randall would continually grind his teeth into what must have been almost dust. It was the most excruciatingly horrible sound you can imagine. It entered through your ears and scraped down your spine, sounding like a quarry full of rocks smashing and grating down in the gnashing foundries of Hell. As it slowed, you could actually hear each separate tooth being ground against the others. He would stare at you while he grinded, his black eyes swimming with nothing.

Halfway through an inane conversation one night, he straightened, as if seized by electricity.

'I've seen wizards,' he implored, eyes burning into me, 'wizards that caper at the feet of waterfalls and great mountains, dancing in the spray, but with their cloaks still as dry as the desert. Some of their auras are light blue, others are dark blue, but a few are black. Black as the night. They look in my direction, but they never speak.'

He paused to grind his teeth against each other with such force that the noise became almost unbearable. 'Maybe their spells aren't strong enough yet?' he asked me.

'That's a distinct possibility,' I said.

'But once they can speak, they'll be able to do lots of things.'

'That's a distinct possibility,' I said.

'Maybe *all* of their auras will turn black . . .'

'Maybe they will,' I said. 'What happens then?'

He looked *at* me for once, not through me as he usually did, and summoned up fear, like water from a well.

'Chaos. Terrible chaos. When the doors open between this world and another, nothing is for certain. Can you buy me a packet of smokes?'

Randall became a surreal fixture, ghosting in when I least expected him and, with each new deluded sermon on the mount, slowly acclimatising me to the Lovecraftian carny vibes I was beginning to absorb. Yet there was one night, several months after I started, where the fluorescent Dali painting I had entered solidified itself fully in my mind.

I'd spent the day – a Friday – at uni sleepily learning about jihadists and revolutions and was seriously struggling to stay awake, even as I rushed, yawning, across the forecourt toward the store, where a group of young kids were holding a bag of goon above their heads, taking turns squirting the golden wine into their mouths.

They held it high, like a shining trophy, glaring distrustfully at anyone who stepped too close. I went inside and was greeted with an angry-looking Ralph, who was raising his watch and staring at it with a look of supreme distaste buttressed by a raft of melodramatic sighs. I was about to apologise for my ninety seconds of tardiness but was drowned out by sudden cries of anguish and tortured moans from just outside the store. I turned around to see the remains of the bag splayed out like a silver starfish as the wine snaked toward the petrol drain. The poor kids were in shock. Several minutes later, after Ralph had huffed off to his car, they formed a funeral line. Their faces solemn and their steps reserved, they marched past the door slowly in file, a requiem for goon.

They soon departed, but were quickly replaced by a group of drunken middle-aged women on a hen's night, gorging them-selves on a feast of imitation-chicken burgers and firing endless party poppers at me so I was covered in long multicoloured streamers, the smell of gunpowder drifting down the aisles. At about 1 am, as I was downing my second coffee, a short, shirtless man with ludicrously hairy shoulders walked into the store. A glint of metal caught the corner of my eye. He was wearing a massive gold cross on the end of an oversized gold chain. The cross was almost the size of his head, bouncing off his gut as he strutted around the store. He bought a jar of Vegemite.

The store began to fill with clubbers; your stock-standard coming-up-on-pills, in-need-of-chewing-gum, blissfully happy, talkative clubbers. Soon the place was buzzing with their pleasant chemical chatter. While I was busy scanning their Extras, Mount Franklins and Red Bulls, an old Italian man had entered, looking hopelessly at the battery cabinet, obviously wanting me to help him with a car battery. As I was too busy serving, I motioned to him that I'd help him in a minute. He gave me a funny look, and then held up his finger, as if to say, 'Ah, I see!' He then began to peel off his shirt, pasty white rolls of fat appearing as even the blissful ravers shuddered. Stretching from his waist to his armpit was a brutally fresh red scar. 'Cancer,' he said, smiling like a loon, his shirt, and everyone else's sense of reality, left forgotten on the ground.

After getting rid of him, I began to settle back into the groove of serving my overly happy customers, before my attention suddenly became diverted by two beaming goths who had walked through the door. Complete with their rings of eye shadow and matching Harry Potter backpacks, they were a spawned combination of anarcho-punks and inquisitive toddlers. But these guys were different. And the difference with these goths was rather overt. It was material, furry and crawling toward me.

A fully grown, rather curious rat – intent on making as many new friends as possible – was scuttling up the middle aisle, pausing every so often to sniff the air and glance back at his giggling gothic masters for reassurance. And then, an unzipping sound, as his drug-affected masters gave the gift of freedom to two more rats from their portable Chamber of Secrets.

I began to ask them what the hell they were doing when a minuscule woman belied her graceful semblance to release the loudest scream I've yet heard in my time here on Earth, piercing everyone's eardrums and alerting them to the rats' presence. The ravers wanted to pick them up and cuddle them, while the woman had already torn off into the night. The goths took advantage of this hubbub to make their escape, laughing that crazed cackle I was beginning to know well. The ravers took the rats with them and went clubbing, or to the RSPCA, I never found out where.

As the clubbers began to dwindle and the night grew long, a new sort of customer appeared: mangy men with handrolled cigarettes tucked behind their ears. Grifters, the lot of them. One of them stood in front of me, randomly twitching while emptying out fragments of a $50 note from his wallet – a jigsaw piece at a time – and then spent several minutes piecing them together. The head of Australia's first published Aboriginal author, David Unaipon, slowly materialised on the counter's promo mat as the note began to take form. When it was finished, he chuckled with joy and asked for sticky tape and two packets of Winnie Reds.

As I tried unsuccessfully to feed the sticky-taped note into the safe's note reader, a man rolled into the store on a wheelchair. He had long wavy hair that fell to his waist in massive clumps. He ignored my greeting, as his eyes were focused on our ice cream freezer and they didn't leave it until he was holding what was seemingly the sole reason for his visit – a Bubble O'Bill. He wheeled to the counter and chucked two coins at me as he ripped the wrapper off and almost inhaled the pastel-coloured bandit. Before I could blink, he'd returned with another. He wheeled outside to finish this one at a more leisurely pace as I went to stock the fridge. But as soon as I got halfway there, the doorbell rang, and as I unlocked the door, he rolled in again, straight to the ice cream freezer to grab another. As he approached me to pay for it, I joked with him: 'A bit hungry, mate?' Again, he said nothing, his eyes glued to his icy treasure.

He paid for it and left. This time I waited, and sure enough, he rolled in again, and I got ready to tell him this was all getting a bit ridiculous, when he raised his head and howled a ninja war cry, flinging a projectile at me.

Instinctively, I stuck my hand out, trying to catch the object, but missed it. In retrospect, I'm glad I did. On later inspection, I found the object was a homemade 'ninja star': ice cream sticks stuck barely together with chewed wads of gum from Bill's famous nose.

I looked back at the man, ready to ask him what the hell he thought he was doing, when he let loose another cry, twisting his massive arms in a Bruce Lee–style fighting stance, before rolling out of the store, cackling to himself.

This all happened in one night. You might be wondering why I didn't call the cops. Well, what would I say? 'Excuse me, but there's a man in a wheelchair who has just thrown some paddle pop sticks at me, please send help.' Or, 'Two anarchist Harry Potter fans high on MDMA have abandoned several rats in my store, please send officers and a pest removal company.' I didn't even bother telling my bosses; they certainly weren't going to hire a security guard to protect me against disabled ninjas. So I grinned – well, cringed – and began to get used to it.

CHAPTER 5

GUMBLETANIA

As technology rushes ever onward, with glinting exabytes of data raining down from a churning orbital soup of satellites as we hold proton demolition derbies 100 metres underground in a 27-kilometre long donut, sometimes it's nice to sit back and think about the tiniest pillars of our existence.

As chemistry developed, atoms were discovered as the smallest matter in existence. These were soon followed by protons, neutrons and electrons, which, if viewed at scale, were each only a basketball floating inside the vast emptiness of an atom twice the size of planet Earth.

Such scale is unfathomable to the human mind, yet scientists soon swung their keening axe again to find the tiniest of the tiny: quarks, which leads us, somewhat tenuously, to quantum foam, which is conceptualised as the foundation of the very fabric of the universe itself.

It was at this level that the madness of the Gumbletons was beginning to permeate me.

Gumbletosis. It was virulent, pathogenic and all-encompassing. I became watchful when out in public. I'd be on the train looking nervously at the guy holding the VB stubbie, or standing on the station platform staring worriedly at the dude with the long blond greasy hair. Normal laughter from schoolgirls sitting across from me would morph in my head into a crazed cackle, making me jump nearly a foot in the air and earning me suspicious glares.

Even at home, Gumbletania never left me. I began to dread the night's coming tribulations, wondering if my shift would be doe-eyed and wrapped in silken shawls, or if she'd pull up in a rattling, spray-painted van with blacked-out windows and blown speakers chortling Limp Bizkit – leaping at me from the sliding door with a jagged machete and a grinning luchador mask. I'd lie in bed at 10 pm, trying to read but my nervous tic making it nearly impossible. Instead, I'd glance nervously at the red minutes ticking over toward show time. Moby would sit near me and stare inquisitively at my newfound anxiety.

Moby is a massive cappuccino-coloured Confucius-like rag doll and is about 50 per cent human, 30 per cent dog and about 20 per cent cat, give or take. I have an excellent relationship with him. Cats have hearing and sight that is at least six times better than ours and are blessed with over 100 vocal cords. Moby would use these gifts to perfection. He always knew my mood, and when I was worried, he would unfailingly ask me what was wrong. He lay there this night, and meowed imploringly at me, gently pawing my face trying to figure out why I was so unsettled.

As I reached the three-month mark of my employment, I realised I had to make a decision and stick by it: quit now and try and find a normal job, or continue with this circus and see where it took me. But I knew I needed money to start adulting, and it couldn't be a job that clashed with my uni lectures and tutorials. So, for my sins, I chose the circus.

Two weekends later I was staring out at the forecourt when I noticed a massive black and yellow shape moving quickly toward me. As it came closer, I saw it was a guy dressed as a giant bee. I unlocked the door and he rushed past me, buzzing audibly up and down the aisles, moving his arms as makeshift wings. After several minutes he finally approached me – with, I shit you not, an already half-finished squeezable tube of honey – and we had a chat.

He explained that he had, in a moment of drunken revelry, stolen the impressively elaborate costume from his mother, who worked for Diabetes Australia and had the 'Diabetes Bee' suit on loan. After squeezing a long amber snake of honey into his open mouth, the bee told me that he would find me some customers. He ran out with the unpaid honey onto the main road just after the roundabout, where he stood, blocking the traffic, flinging honey at their windscreens and directing them into the servo. Motorist after bewildered motorist appeared, thinking there was some highly unorthodox promotional sale going on. I had to spend five minutes assuring one very irate elderly man that it had nothing to do with the servo; it was just a random lunatic in a bee suit.

The Gumbletons were here to stay. And so, ostensibly, was I, slowly becoming one of them myself. I learned to get into that weird headspace every night, that kind of flippant, uncaring facade where anything was fine by me: sit on the ice cream freezer and recite poetry, dance around the aisles like a drunken Cossack, lie on the floor moaning and covered in *Sunday Herald Suns* – I didn't care. Surely by doing this, I would convince them all that freaking me out would not only prove exceedingly difficult, but wouldn't be any fun either. And it worked. For a while. Most of them did their little dance, got bored and drifted away.

But then something terrible happened.

Looking back, I blame George Orwell. I'd just finished reading *Down and Out in Paris and London*, documenting the hardscrabble lives of those less fortunate in the early 1900s, and something empathetic bloomed in me: I actually started caring about the Gumbletons.

Everyone has a story, and as I began to listen, I learnt theirs: no one becomes a Gumbleton all at once. There were so many seemingly inconsequential twists of fate, dollops of unfairness, and consecutive bad decisions in a person's life, each of them tilting the axis it spun on just a smidgeon more off-kilter until, one day, they wobbled too much and their orbit veered completely out of control, careening away from the nourishment of the sun through vast wastes of interstellar space to now circle a leering black hole. Something about that journey resonated deeply, as I realised a Gumbleton could be anyone, including me.

So I fed them free pies and sausage rolls from the warmer, stopping all pretence of work as the long hours of the night crept toward day. And when you let one in, providing shelter from the harsh climes of a Melbourne winter, you let them all in.

While I didn't get any much-needed uni work done, my conscience was somehow clear. As the weeks crawled by and winter deepened, the crowds swelled. Alcoholics, addicts, the mentally ill and the homeless mingled together in my retail utopia, rubbing their hands in the colourful warmth, feeding me tall tales from their many internal adventures as I nodded, wide-eyed and eager to learn about life from the third side of the coin. I learnt many useless but interesting things.

Like, exactly what time unemployment benefits are paid into accounts. They all had it down to the exact minute, their watches synchronised to Centrelink's as they rushed to the ATM so they could buy their smokes. I also learned what time the connections to the bank lines went down for their weekly update. This was back when not everything was fully computerised and I continually wondered to myself why so many people with Commonwealth Bank cards seemed to emerge on a cold Sunday night to waste money on a feast of overpriced crap. Little did I know that there was no link to the bank and all of these goods were being provided gratis, care of my servo. I figured it out, eventually, but no one said anything, so the store was always uncharacteristically busy on Sunday nights, just after midnight.

But there was so much more of an education to be had. Thanks to my merry band of miscreants, I now know the answers to the following questions: What happens when an old homeless man smokes nine cigarettes at once? (He coughs up half a lung, and his trachea, then falls onto the – now cracked – ice cream freezer.) What happens when two wild-eyed philosophers disagree on the very definition of third-dimensional reality? (They each seek to prove their respective theories using Doritos, salsa dip and *Simpsons* figurines.) And what happens when someone with serious mental health issues replaces their anti-psychotic medication with liberal amounts of methamphetamine? (Let's just say it took a while to clean up.)

It was about four months into my servo journey, and one thing I noticed was that even outside of work, this coalition of schemers seemed to sense I was becoming one of them. It was as if I emitted some strange empathic pheromone proclaiming my sympathies for all the downtrodden and eccentric ne'er-do-wells. That week I was walking down Werribee's main street with my mum and grandma, heading to a restaurant for lunch. Every 100 metres or so I'd hear a gargled cry, and turn to see one of my freaks, festering in the shadows. Ignoring them did no good: they simply followed me and raised their lacerated voices until I had no choice but to stop. By the time we got to the restaurant, my mother suggested we take a different route back to the car. It didn't seem to help much; a couple of them still managed to wobble across our path.

A few weeks later I took my younger brother into Centrelink to help sort out his youth allowance. Big mistake. It was a gathering of Gumbletons heretofore unseen: writhing, cackling chaos on an unimaginable scale. I'd found the nest.

It was the same on the train to uni. As I had to travel to Burwood, in Melbourne's far east, I had plenty of time to acquaint myself with being an unwilling magnet for the unhinged.

One day, as the train rumbled to a stop in Footscray, a droopy blond mystic-looking fellow hopped on. Ignoring the mostly empty carriage, he staggered his way toward me, an acoustic guitar slung over his shoulder. He had a powerful miasma that surrounded him like a force field. It was part bourbon, part weed, with a pinch of spew. He sat down opposite me, starting on the six-pack of bourbon and Coke he had with him. He grinned at me and downed the first two cans as quickly as I've ever seen, burping like a ship's foghorn before scratching his yellow beard and grinning at me again.

'Ya like *music*, champ?'

'I don't mind it,' I said noncommittally.

'Howzabout I plllaay a *SONG* for yaaa?'

'Go for it.'

He started drunkenly strumming the strings, repeating the same two chords again and again. As my attention waned, and my amygdala quietened its distress responses, I turned to look out the window when a horrible wailing ripped my head back to the source of the noise. It was his chorus, an aural nail screeching up a forsaken blackboard, held by a demon clambering up from the charred pits below. At the next stop I rapidly farewelled him, pretending to get off, but instead ran onto the next carriage.

This move paid off, for it brought Joe into my life, albeit for only ten minutes. A can of VB in one hand, lit cigarette in the other, his eyes protruded from his unshaven face, darting nervously left to right every few seconds, searching for danger, or perhaps authority figures. His flimsy tattooed arms seemed to be an extension of his green, mangy singlet. Upon entering the carriage, he, like our musician friend, ignored row upon row of mostly empty seats and made his way over to mine.

'*G'day champ!*' he boomed, as he draped himself across the seats opposite, taking a final drag on his ciggie followed by a long VB chaser as he appraised me in full. 'What's fucken news?' Upon learning my name (I told him it was Memphis), he continued the conversation, seemingly not too sure about his own.

By the time he'd remembered his name, I'd been given a fairly detailed summary of Joe's views on abortion ('Punch her in the guts 'n' save a fucken packet!'), fare evasion ('Ask me for a fucken ticket and I'll knock ya inta next week, ya suited cunt!'), and the always topical question on whether the woman's breasts two rows of seats behind us were actually real ('Maybe real, maybe not. All I can tell ya is it's Ray's Tent City down here!' – said as he gestured to the bulge in his jeans).

Before we went our separate ways, Joe, glancing across the graffiti littering the walls of the carriage, lamented to me, fresh cigarette in hand and his feet sprawled across several seats, 'Why can't these little shits simply follow the rules like everyone else?'

So, it was official, I was Brother Teresa Dark Edition™. I suppose in one way I didn't really mind: some writers spent their whole lives looking for vivid characters and crazy plotlines. Here I was being given pre-assembled misanthropes, free pies and a steady salary. Hunter S. Thompson, eat your heart out.

Still, after another few weeks on the train to uni, giving off my strange freak-pheromones, I finally grew up and got my licence and a car, all within two days. It's funny what motivation will do.

And while the Gumbletons were proving fascinating – so much so that I'd started scribbling down their exploits onto register receipt rolls between customers – my actual uni work was piling up, and no matter how hard I tried to re-engage myself, I found my denizens of night way more interesting than Mao's Cultural Revolution.

One thing I noticed during these months were those tiny, transient snatches of madness. An action or feature just a little out of the ordinary, so quick you'd blink and miss it. Like a tough-looking general dressed in full uniform with epaulets and medals who handed me a discount docket and incidentally touched me with the most ridiculously soft hands. They were like velvet. He must have moisturised them thrice daily.

Or a corpulent dude with a Cheshire-Cat grin wearing artfully applied rainbow body paint, a toga and an inflatable giraffe pool ring, emerging from the rain-soaked night and dripping colourful splotches on the floor as he sprinted straight for the toiletries section, before bolting back to the register with a box of ribbed condoms and a manic grin.

One guy came in three Saturday nights in a row and would only mime what he wanted, complete with exaggerated frowns when I misinterpreted him. One woman would only ever fill up with $55.55 worth of Premium Unleaded on pump two. No exceptions. Even if every other pump but that one was free, she would wait for it. And then there was a hairy miscreant, who, after paying for his petrol, used to enjoy jumping into waiting cars and horrifying the strangers inside.

The most constant one-trick pony was this guy who'd walk in every Sunday night and ask: 'Got stuck with the nightshift again, mate?' Once a week, every week, for the better part of three years. In exactly the same tone and pitch. With the same lifeless grin on his face that never touched his eyes. I think he scared me more than anyone.

With no sleep or sanity, but at least some actual money in the bank, I was at my mate Hassan's local Indian restaurant for a chicken biryani. It was a slow lunch day and he asked me what

I'd been up to, raising his eyebrows when I told him I'd recently got a job working nights in a servo.

He grabbed some garlic naan and two Kingfisher beers from the fridge and ventured over to my table, sitting down and facing me with wide and suddenly serious eyes.

'What's it like so far?'

'Like a desolate alien moon in a galaxy far, far away.'

He grinned, ripping off a bit of naan. 'That crazy, huh?'

'I'm starting to feel like a figment of my own imagination.'

That got me a knowing grin.

'I understand you completely. That's why I got into restaurants. I couldn't take the craziness anymore. All the weirdos, the rudeness, racism, the drive-offs and violence . . . And the questioning inside your head if you were slowly becoming one of them.'

'Yes! Wait, you've worked in servos too?'

'Yep. Twelve long years, brother. Man, the things I've seen . . .'

'I can only imagine.'

'You know, I delivered a baby once.'

'*What?*'

'On the back seat of a woman's Commodore on the forecourt, in the middle of a freezing night, in St Albans.'

'Wow.'

'I got a black eye for my good deed. And I *still* got in trouble from the boss for not getting all the delivery put away that night. Which servo are you working at?'

I told him and he laughed. It was, of course, the same one he'd worked at for years.

'I know this might sound a bit weird, brother,' he said, leaning back in his chair and gulping down his beer, 'but there's something about that specific site that's . . . not quite right. I worked there for about three years and the customers that show up there ar—'

'Freaks.'

'You got it! It's almost like—'

'A wormhole has opened up nearby somewhere.'

He cracked up laughing. 'Got it in one, brother.'

The bell above the door jingled and a young couple walked in. Hassan got up from the table and gave me an appraising look.

'It's a dark circus, that place. Don't stay there too long, or we might lose you forever, Dave. Get out when you can. I only started this place to get away from servos. People are much happier when you serve them food. And the dark parade of freaks are nowhere to be seen.'

Back behind the counter, with Hassan's words ringing in my ears, I realised I *was* watching a circus, but I was smack-bang in the middle of it; the ringleader himself, attempting to mould all the chaos into some sort of show for an eager, invisible audience. I knew I couldn't actually control it, only offer encouragement and try to diffuse madness as it began to bloom.

Like any circus, there are clowns, in fact there's an abundance of them: a multitude of brightly coloured fools in baggy clothes dancing around the aisles in endless kinds of stupor, providing hours of mindless entertainment. There are lions also, ones that drive Commodores and smoke Winnie Blues, and take great skill and patience to tame. Occasionally their roar will fill the aisles and make the audience shudder, but usually they behave.

And then there are the freaks, which deep down every circus needs, so that we can 'ooh' and 'aah' and feel better about ourselves. As you now know, they come in all shapes and sizes: from psychotic warrior behemoths, to emaciated mopers, to all the addled acrobats, walking a precarious tightrope by discovering how many pills they can consume before they pitch themselves into the net below.

But by some weird twist of fate, it wasn't the freaks or the lions that were caged. Instead I was the one hemmed in by a glass door and anti-jump protection wire, free to observe the greatest show of suburbia in all its surrealism.

Sometimes the show was a cliff hanger. One morning, in early December 2003, having survived almost a full year of servitude, I was returning to the store from the back fridge when I heard the approaching drone of a motor, followed by the signature squeal of locking brakes and yelping tyres. Dropping the milk crates I was carrying, I rushed to the window to see a late-model silver ute slide through the wrong side of the roundabout. With a *donk*, it mounted the kerb and headed cross-country, fishtailing across the grass and gravel, sending pebbles leaping from its tyres as it headed straight for my parked car.

At the last second the driver swerved, cutting across the servo's front driveway, smashing the bollard, exit light and a tap in the process. The car finally stopped centimetres from the price board. Almost instantly, the guy leapt out. He tried to run, but he was seemingly too drunk, so he fell down on the forecourt, where he flapped like a landed fish in the gathering pool of water from the burst pipe connected to the crushed tap.

This kind of thing happened a lot at my servo. One foggy morning I heard the usual screech of skidding tyres and approached the window to look. I saw the ghost of a white Hyundai 4WD scythe through the roundabout and then plough through a stone wall and wooden display-home sign. As the horn from the wounded car wailed into the dawn, I sighed a familiar sigh and called an ambulance. But somehow, again, the driver was unhurt. He climbed out of the car and ran off into the murk.

But I've seen several people die over the years, their cars rocketing into a power pole in a carnival of flames. So many just didn't seem to care. They were too far gone. Sullen. Medicated. Sloshed. Flying. Tanked. So many lost souls, solicitous to destruction. Black eyes that didn't care for consequence until it surrounded them like a snake, suffocating them with its newfound reality.

Some, though, had run out of things to lose long ago.

It was early Christmas morning and I was squirting sauce onto a stale sausage roll between drunks when an angry cry made me miss my target, coating the floor in a vein of red condiment.

'*Vampires!*'

Randall burst into the store in a state of ecclesiastical fervour.

'Drinking the blood of Christ is a *symbol*! It was never meant *literally*,' he scolded me, his eyes smouldering like black coals. 'Madness abounds when morals fall away. Do you know madness?'

I remember thinking to myself I had a fair idea.

'True madness,' he continued, 'the kind where everything falls away, and you're left floating on the void, wondering if you'll fall up into the teeth of roaring nothing. I hear voices . . . sometimes. They don't say anything though.' He smiled absent-mindedly, looking as though he was remembering, as his eyes got that glazed, Nick Nolte look about them. For some reason he smelt strongly of sawdust.

He stayed like that for twenty-five minutes, mumbling incoherently, as the Christmas clubbers came in with their papier-mâché party hats and inebriated sense of festivity, pulling crackers and giggling at the bad jokes contained therein.

One of them staggered up to Randall, waving a bright blue cracker in his face. 'Hey buddy, *pull my finger*!' Randall stared past him, deep into his other place, and started grinding his teeth to dust. Rapidly the guy lost his nerve and crept out of the store, tail between his legs as he shot an anxious WTF look back in Randall's direction.

Randall continued with his trance for another half-hour, sometimes here, sometimes not, rocking back and forth on his heels as he lashed the young revellers with tales from his own bible. He was a Grinch, in both action and appearance as he laid down the fire and brimstone, before floating away again to his netherworld.

After he'd finally finished, I bought a pack of PJ Gold 30s and placed them on the counter.

'Merry Christmas, you crazy fuck.'

CHAPTER 6

INMATES

Servos have a high turnover of staff. Go into any place often enough and you'll notice a new, nervous face that says everything it's supposed to, pestering you to buy 2-for-1 promo specials, stapling your receipts together, and smiling like a Stepford Wife. Remember that face because it won't last long. Whether they can't handle the freaks, have too many drive-offs, or they're just too damn slow, they'll be gone by the end of the month. The ones that stay are the crazy ones.

Ralph was crazy. It took me about three and a half months to really confirm it. Maybe not a running-around-with-a-beheaded-rat-and-blood-dripping-down-your-chin-while-dancing-the-macarena crazy, but more a slightly-unhinged-well-that's-a-bit-out-of-the-ordinary-he-really-needs-to-get-out-more crazy. No, that still doesn't describe it. You know those old ads, years ago, for the tax people? The ones that find a minuscule tax deduction for their client, check to see if anyone's looking and then pump their fist in the air? That's Ralph. To a tee.

Ralph loved our servo. The place seemed every square inch of his life. Of all the console operators I've seen, he was by far the most committed to his craft. You'd be getting your till ready when he'd burst away from the console, scarlet-faced, out the auto-doors and onto a busy forecourt, weaving between shunting cars as he demanded retribution from a customer that had short-changed him 5 cents. One stinking hot night in early 2004, I walked in to start my shift and I realised he wasn't his sprightly self, but was moping around like a grounded child. After asking him what was wrong, he told me that he wasn't happy with the till I had left him the previous week. It didn't have enough 50-cent coins. He told me he expected better of me in the future.

He'd pop in several times during the week, not filling up his car, just . . . checking on things: usually his, and everyone else's promo sales figures. He could never wait the extra couple of days to find out how he went. Whenever I was rostered after him and I'd come in at midnight to start work, he'd always propose extra tasks I could consider doing that night, and would then proceed to stay until 1.30 am, reading passages out of *Woman's Day* and glaring disapprovingly at entering customers. He'd just linger, like a bin chicken in a park, picking at this or that, slowly driving you insane. He once stood on the other side of the console, reciting the entire Tasmanian batting card after a recent one-day match, minute after excruciating minute, while I was trying to serve customers.

A born sycophant, he'd heap praise on the boss's unpalatable requests, completing all the easy, prominent jobs that would be noticed first, while directing others toward the shit-shovelling at the first possible opportunity. He'd hover around you as you worked, helpfully suggesting all the ways in which you could perform the job better. More change in the till. Products could be straighter and barcodes more aligned to the left of the shelf. That kind of thing. Of course, he didn't have any actual authority over any of us, at least outside of his own mind; he was the same lowly 'customer service operator' as the rest of us. He just felt the need to direct people.

Ralph applied for the 2IC (second in charge) position three times in the years I was there and was denied each time, despite being the most senior and longest-serving applicant for the position. On the final occasion, he was overlooked in preference for a woman who had been around all of fifteen minutes. Although we knew he'd never get the position, the store breathed a collective sigh of relief each time we found out Ralph would have no power over any of us.

Despite all of this, a part of me loved Ralph. There were many reasons. One being that despite appearances he was a selfless paternal figure to the newer workers in the store. For my first few months, he was my mentor and protector. For the hour he'd sometimes hang around on my shift, he'd stare down psychos, trying to shield me, the callow newbie, hiding wide-eyed behind him. Every so often, he would, in his own time, pop into a nearby store and help the person working put away the delivery or stock the fridge.

But the main reason I liked Ralph was that he was simply good value. A servo can be a dull, monotonous place sometimes and Ralph made it a lot less beige. A few nights into the new year, I entered the store and saw no one behind the counter. As I came further, I found him sitting on the floor, counting the change in the change machine. His shoes and socks had vanished and he was sitting there like Buddha, legs crossed in a kind of lotus position, bemoaning the fact that his podiatrist was younger than him.

Sometimes after reading the print-out of his shift report and noticing an unexplained variance, Ralph would stiffen as if shot, the colour draining from his face as he began to seethe with something halfway between fury and self-loathing. He'd run his hands through his hair, and pace the console with a manic energy, rifling through receipts and vouchers like an addict searching for his lost stash. If he didn't find the discrepancy, I knew I was in for a long night. On other occasions, after reading the print-out and seeing his $0.00 variance, he would pump his fist in the air and let out a victory cry. He'd then comb the diary for days after, searching for praise from above that, sadly, never came.

One midnight, as I got in, I noticed Ralph had left the weekly newsletter from our retail area manager in a prominent position, solely so I could see that he was head of the weekly footy tipping competition. I took the newsletter from the back room and began to leaf through it casually, in plain view, studying his reaction from the corner of my eye. As I remained silent, he became fidgety, restless. He skulked around the store, attending to odd tasks that didn't need doing, waiting for some form of acknowledgement. I smiled to myself, enjoying the moment, though I probably shouldn't have.

'Go to the footy this week?' I asked him.

'No.'

'Yeah, me neither.'

After going to the toilet, I came back, counted my till and started serving customers, noticing the newsletter had disappeared. Thinking he'd given up, I turned to get a customer a packet of smokes and came face to face with the newsletter itself, Ralph's shaking finger displaying the source of his uncontained pride.

'See this week's footy tipping?'

'Well, I can now,' I said as I retreated from the newsletter, so close it blocked my line of vision completely. 'I see you're on top.'

'Just lucky, I suppose,' he said, which was clearly my cue to congratulate him.

Instead, I sighed and grabbed the Winnie Golds, spinning back to the relief of customers.

As the months dragged on, my shifts were chopped and changed, and I saw less of Ralph, but got to meet many more of my fellow employees. Some women I've worked with are tougher than any man I've ever known. Service stations seem to breed them: tenacious, bulldog-tough and ready-for-anything women, who would bluster about the console, leaving you in no doubt as to who was in charge.

But they also had the most amazing warmth and sincerity. Cross them and they'll get revenge. Get to know them and they'll be a great friend. Nellie was one of these women. She was the stayer of our servo, having been there seventeen years, from the day it opened; which was more than twice as long as anyone else. She was bossy, stubborn, hilarious and friendly.

Critical of incompetence but constantly self-deprecating. Like Ralph, she was simply good value. At first, though, she was downright intimidating.

'I'm not doing Sundays anymore after this week, so you'll be having Nellie from now on,' Ralph told me one night as he was closing off his till.

'Okay, cool. What's Nellie like?'

'Umm, it's . . . an experience.'

The following weekend came and I was somewhat apprehensive as I showed up to work that balmy summer's night. As I walked into the store, a battle was raging. Behind the counter stood a tiny brunette woman who was staring up defiantly into the angry eyes of a bald, muscle-clad psycho, his nostrils flared as he hurled every vile threat he could think of in her direction. She spoke with a thick Balkan accent.

'I'm sorry, sir, but I told you that I cannot do that.'

'I wanna speak to your manager.'

'I'm the manager.'

'Well, why doesn't it say it on your name badge?'

'Cost-cutting.'

'Listen to me, you fucken wog bitch, I just wanna get a different pack of smokes. So just swap 'em and I'll be on my way.'

'No, you listen. I just want to eat my pear in peace and get back to counting my till. So you can just be quiet and leave the store and I won't be calling the police for the threats you've made against me.'

The bald man pelted his smokes at her head in disgust, which she caught with snake-like reflexes and smiled, before picking up the phone to ring the police. Sensing trouble he split, leaving me staring wide-eyed at the crazy woman holding the packet of Winfield Golds with a crazy look of triumph.

'Do you smoke, David?' she said, following him with her eyes to his car.

'Umm . . . no.'

'Good boy.'

The psycho's yellow Monaro growled as he revved and revved it, then, tyres screaming, shot off angrily into the night.

'I smoke. Too much. Winfield Reds. The strongest and quickest path to cancer. Do you know why?'

'No . . . ?'

'This place. It seeps into you and it stays there. In your bones, David. That's where they all live. All the angry meatheads. All the rudeness and the racism. All the crazies. Until, one day, you realise that you've become a crazy, too.'

She turned to examine me.

'Why are you here, David? What possessed you to join this circus?'

'Well, I'm at uni and needed a job, and I suppose it was an easy job to get.'

'There's a reason for that, David.'

'I think I'm starting to understand that.'

She gave me another measuring gaze.

'You will. In time. You will learn lots of things if you stay here. And you will change, David. For good or bad I don't know. But I know that you will change.'

She brightened, suddenly.

'Well, I'm free again from the salt mines. Your till is ready, do you need to count it before I go? No? You trust me, David? That's a good start. I just met you but I think I trust you, too.

Don't let them boss you around. Stand up for yourself. I'll see you next week, David. Goodnight.'

Not long after I met her, Nellie divorced her second husband, after discovering he no longer found her attractive, preferring young hairless men instead. While this may have bothered some people, Nellie didn't mind, she just got on with things, working ridiculously long hours at the servo to pay off the divorce settlement. She got a second job as a prison guard, splitting her time between the big jail and the little brightly coloured one. The thing was, the more time she spent at the prison, the more she realised how nice a working environment it was compared to our servo.

'It really is very similar, David,' she said one night, leafing through *Woman's Day* before tossing it onto the counter in disgust. 'Why do I read this toxic horseshit!? Anyway, like I was saying, the prison is maximum security, David. Vast, cold and foreboding, endless cameras picking up everything you do, just like here. The people who work there are crazy and depressed, also just like here. I really do feel quite at home. And even though it's full of the worst of the worst, rapists, thugs, paedophiles and murderers, you know what, David? I prefer it there. And do you know why?'

'Umm, no . . .' I was still grappling with the notion that there were so many similarities between the servo and a maximum-security jail.

'Because even though these brutes are locked up for decades, none of them come close to whingeing as much as the customers do here. There's more humanity in one small corner of that prison than there has been in seventeen years of this forsaken hellhole. Give me the murderers any day of the week, David.'

I gestured toward a waiting customer, but Nellie wasn't finished yet.

'Plus, at the prison, David, the prisoners are the ones who are locked up. *Here*,' she gestured to the locked coded door and its aluminium-framed cage with anti-jump protection wire strung taut over the counter, 'here, *we* are the inmates.'

Nellie had this way of cutting through people's crap and getting straight to the heart of the matter. She wasn't *completely* rude, but she could be . . . brusque. Work in a busy servo for over seventeen years and it's kind of inevitable. She wouldn't bother with the 'Hello' or 'How are you?' She went straight for the 'Pump?' or 'What do you want?' To the people who expected to be treated like esteemed royalty when they walked through those doors, she was a blast of frigid air that snapped them out of their delusion.

She had a great way of dealing with customer complaints, too. One Sunday morning, as I was finishing a long, crazy grave-yard shift, a customer approached the side of the queue and complained that his custard donut had no custard in it. Nellie offered him a replacement, but within a minute he returned with the same complaint. This went on for another few rounds – return, replace, still no custard – until Nellie broke the stalemate.

Ignoring the waiting line of customers, she marched over to the donuts and grabbed an armful, bringing them back to the counter and ripping them apart with her bare hands, searching for the elusive custard. Seven donuts later, the counter a jagged landscape of dough, she discovered a small speck, which she held up to the light and showed the by then nervous man with a kind of elated pride.

'There's your custard, sir! I knew it was in there somewhere.'

And with a sweep of her arm, she launched it all to the floor, shouting '*Next!*'

Nellie also took a keen interest in the welfare of the flora surrounding the servo. She would often liberate plants she felt were being neglected, digging them up in the dead of night and taking them home to give them a better life in her backyard, where they'd be treated with attentive loving care.

She'd also *bring* lots of stuff to work as well. As I trudged in on Sunday midnights, she always had little edible gifts for me.

'You don't look well, David. This place is changing you. At least we can keep your vitamins up.' She'd then hand me a container with sliced pieces of obscure new fruits. One night it would be lychees, the next it was kumquats.

The thing you had to remember, though, was she'd always crack the shits if people used her stuff without telling her, leaving massive angry three-page notes in the diary, demanding retribution. Scrawled in red pen and pressed so hard you could feel the grooves on the other side of the page, they always ended with a demand for the offender to publicly identify themselves:

PLACE YOUR NAME HERE: _____

I spent the next few weeks and months marvelling at Nellie's unique customer service skills, while at the same time trying to extract method from the small snatches of her madness I saw on shift changeover every weekend. But little did I know I was soon to have another teacher.

Donna was, well . . . Donna. There's no other way to describe her, although, I'm at least going to try. And to do so, I must first describe a process familiar to us all, involving a rather well-known central nervous system depressant.

Low-level alcohol intoxication is the process whereby ethanol, alcohol's active ingredient, leads to a noticeable reduction of inhibitions at a cognitive level, whereby the subject will often act in a forthright manner, doing and saying things they would normally keep censored, for fear of awkwardness or social embarrassment.

Due to the many vagaries of human physiology and behaviour, scientists find it difficult to gauge at what actual point the ethanol is lowering these imaginary gates of inhibition and releasing the ribald hounds of intoxication. Often, it is said to be necessary to observe the subject's behaviour while they are sober, so as to establish a baseline.

I observed Donna in many differing situations, and as a test subject she was confounding, solely because lack of inhibition seemed to comprise the actual bedrock of her psyche. The woman was as mad as a bag full of cats.

She would *always* speak her mind, no matter what ribald profanation backstroked through it. Although 'speak' isn't exactly the right word. It was more a twisted cache dump of ammo, untrammelled by the safeguards of decorum or societal expectation, used to batter whoever was on the receiving end into a state of subservience through sheer and utter bewilderment.

In late January 2004, we were running low on staff and I noticed her name on the roster for the following week. Curious, I asked Ralph the next night, as he was counting the change in his till, if he'd ever worked with Donna. The coins were clattering into the slots as we talked, in Ralph's slow deliberate style, but as soon as Donna was mentioned, the pace of the falling silver became erratic, with one or two missing the till altogether and rolling along the dirty black linoleum toward me, punctuated by Ralph's mysterious silence. As the quiet grew, the 20-cent piece dropped to a halt at my toes. I picked it up and handed it back to him, noticing he'd stopped counting and was trying to form a sentence.

'Donna is hard to work with, sometimes.'

'Why's that?'

'Well, she's just a bit, well, crazy sometimes, well . . . *all the time.*'

'What does she do?'

'Everything she's not supposed to.'

'How does she still have a job here then?'

'You know, I'm not really sure . . . I've never really thought about that.' Then, evidently pulling himself together, he added rather too cheerfully, 'You'll meet her soon enough, then you can see for yourself.'

I was intrigued. After all, it was Ralph who had assured me on my first shift that the majority of Werribee's night-faring population were all 'pretty normal' people, so I wondered what would qualify in his bespectacled eyes as 'a bit crazy'. A week later, I was given a front-row seat.

As I entered the store that night, a long line of impatient customers stood cross-armed, cradling their Red Bulls and bags of Twisties as they waited for the transaction that was holding them up to be completed. But a transaction, dear reader, is a trade that requires two willing participants.

A young woman was standing at the counter waving a small card around in a high state of annoyance. She was faced by an older, seemingly less-annoyed woman whose mouth was turned up ever so slightly in the suggestion of a smile. In a voice that twanged like a slow, drunken rubber band, she said, 'I dunno what kinda card that is but it ain't gonna work here. Ya can't make strawberry jam outta pig shit, luv, no matter how much sugar ya use.'

'But it's a card from your own damn company!'

'S'not my company. And I've never seen it before. You got another one?'

'But I shouldn't *need* another one!'

'And I shouldn't need to have a piss right now, should I? I shoulda done that an hour ago, but what can ya do? Gimme the card, we'll see if we can work some magic.'

She picked up the card and was about to swipe it when she noticed me rounding the console. She stopped with the card in her hand and turned to inspect me.

'And who do we have here?'

'Hi, my name's D—'

'Are you here to save me from the hordes?'

'Well, I sup—'

'Well, I'm not the one who needs savin'; these drongos need savin' from 'emselves. I'd rather shit in my hands 'n' clap than serve one more of 'em. You ever seen this card before?' she asked as she flung it my way like a frisbee, the card slicing haphazardly through the air, missing my fingers to fly through a small opening and onto the top shelf of the pie warmer.

At this she cackled, turning back to the now speechless young woman, saying, 'Well, maybe it'll work better with a bit of heat, luv! Hah! That reminds me, I need a fag. You're right with these guys, yeah?' she asked, gesturing to the nervous line of waiting customers, each wondering what their own encounter would bring before they visibly relaxed as I answered in the affirmative.

And with that she grabbed a Snickers bar from the shelf, tore it open and tramped for the exit, having lit her cigarette even before she was out the electric doors. A buttoned-down man at the end of the line gaped at her in shock and Donna stopped, hand on her hip as she blew out smoke and sized him up.

'Who pissed in your cornflakes, knackers?'

And with that she exited, past the leaking gas cylinders and into the wafting petrol fumes that rose from the forecourt, chuckling to herself as she headed toward the car wash.

I inhaled and exhaled deeply, before rejoining reality and fishing the now-furious woman's card out of the pie warmer and de-flaking it before running it through the card reader to find out it worked fine after all.

As she prepared to leave, the woman looked me square in the eyes and, trying to dredge up anger, she lowered the pail but brought up only buckets of bewilderment, as many a customer would in the future when dealing with the confounding human hurricane that was Donna.

'I've never in my life experienced anything like that,' she declared.

'Neither have I,' I said, truthfully enough.

I didn't see Donna often after that night, but when I did, I was always reminded who I was working with. She had this somewhat embarrassing habit of – when I was serving a long line of customers on a Sunday morning – grinning and asking me, 'Did ya get a root in the fridge last night?' Once this woman piped up that she didn't think that was appropriate 'for this public environment'. Donna just turned to her and after looking her up and down, laughed and said, 'I don't think *you're* appropriate for this public environment, luv. We'd need a tractor to drag the needle outta *your* arse.'

Donna would sporadically steal items from the store during her shift to brighten her mood. You'd see her grab a chocolate bar and rip it, saying, 'Oops! It's damaged goods now!' before hoeing in, then repeating the process, and offering everyone in the store their own 'damaged' favourite. She lived on a farm a few hours away, but made the trek in several times a week. Considering she had another job (at an opposition servo, no less), plus the farm to run, I could never figure out why she bothered. I think she just relished the cut and thrust of customer warfare.

Some of these customers were actually scared of her. When I covered a shift during the day once, and was out emptying the bins on the forecourt, I heard two tradies standing by the pump, arguing.

'You goin' in?'

'No, I went last time. She's fuckin' loopy.'

'It's your car, mate, and you gotta pay for your petrol,' the other one laughed, as he hopped back into the ute, smiling at his mate skulking toward the store with a hunched back and ice water in his veins.

Customers had good reasons for their concern, though. Once, Donna was working alone at a nearby store when a large stocking-headed man wielding a baseball bat ran in, demanding money. Donna ignored him and continued to serve the startled customers, as if he didn't exist. One can only imagine what went through the young man's head. At the first sign of any hold-up, console operators are, of course, supposed to surrender the entire contents of their tills, their wallets and whatever else takes the robber's fancy.

After several minutes of standing there wishing he could vanish through the cracks in the tiles, he bolted out the door, followed by Donna's delighted cackle, and a 'D'ya want a receipt with that, ya dickhead!'

Instead of being praised for foiling a hold-up, Donna was punished, and in her absence, she was replaced by a guy named Rathindra. I immediately christened him 'Mr Happy' – with good reason, for he was the most cheerful, peaceful soul I've ever met. After only twenty minutes with the guy, you got the very real feeling that you could steal his wife, key his car, lay a hot steaming crap inside his letterbox and then ask him to cover your shift, and he would still smile and accept. Not because he needed the money, but simply because you asked him to. He was just . . . indefatigably agreeable.

All my mates would ask me who that 'weirdly happy Indian dude' was. (For the record, he was Bangladeshi.) Every customer, no matter how belligerently drunk, helplessly stupid, imposs-ibly rude or just damn mean, was to him another chance for a shared joke, a valuable lesson or a comforting word, as he *revelled* in the solid stream of flotsam adrift among the gutters of humanity, truly putting the 'service' in our servo, wide-eyed, helpful and eager.

Nellie didn't trust him.

Night after night Rathindra would stay until 2 am, chat-ting amicably while placating the raging hordes *and* putting away my boxes from the night's delivery. As grateful as I was, I felt his wife and newborn daughter might like to remember what his face looked like. I'd end up having to almost physically threaten him to get him to leave.

One night we'd been talking for only ten minutes after he'd finished when he told me he should probably get going, as his baby daughter was in hospital with some mystery illness. I asked him why he didn't ring me to come in earlier and he said he didn't want to inconvenience me. He once worked a continuous sixteen-hour shift, because the other person didn't show up. He told the retail area manager not to bother looking for someone else, because he wanted to help out.

About two weeks after I met Rathindra, I was leaning sleepily on the console at the start of my shift, monotonously chatting to him, tired beyond belief, having spent the day trying to keep up with essays instead of sleeping, when a searingly bright orange ute pulled into the forecourt. As it rolled in, an ear-splitting '*bahhaahaahaa*' of a sheep, then a neighing of a flustered horse, poured out from its hidden animal horn as it approached the pumps. Out came Stevo, all six-foot-five of him, hairy gut hanging over his equally bright orange fisherman's pants and thick purple woollen socks.

He filled up, came in and got a king-sized slurpee, looking around furtively before reaching slyly into one purple sock to extract a hip flask of vodka to freshen it up along with a Red Bull from the fridge. He stirred it all up, had a sip, then brought it up to the counter, where he took a long weighing look at me.

'You look like a swamp rat.'

'Cheers.'

'Seriously, you look like arse. I think it might be terminal.'

'What?'

'Gumbletosis.'

I sighed. 'You might be right. You were definitely right about the Gumbletons.'

'I know,' he said as he held his card up ready to swipe, but delaying.

'What?'

'Aren't you going to smile and ask me if I want two packets of chewie for three bucks?'

'Don't push it.'

'You know what? You need to get away from this place. At least for a day or two. To bask in a completely Gumbleton-free environment to recalibrate what's left of your soul.'

'Sounds like a plan.'

'Sweet, next Saturday, lock it in, Eddie.'

I handed him his receipt and he looked around the console beseechingly.

'Oi, give us the toilet key, I need to go snap off a grogan.'

'You're not gonna block it, are you? It took me a while to unblock mine the last time you came over.'

'No promises, sweet cheeks.'

Twenty minutes later I heard the signature crow of a rooster, and a banging door, followed by Stevo, almost skipping back with the key, a look of both relief and distaste on his face.

'Poor water flow, skinny pipes. Recipe for disaster. And I sure baked a cake in there. Might wanna get someone to take a look at it.'

He sauntered out to his orange ute and drove away, serenading us with a violently snorting pig.

Rathindra was enthralled. 'What an interesting young man! Are all of your friends like this?'

'No, there'll only ever be one Stevo.'

I managed to get Rathindra to cover my weekend shifts, and a new guy to cover his, and Stevo picked me up at the crack of dawn and we headed down to The Prom, the indescribably beautiful coastal national park at the far southern point of mainland Australia. It took three hours to drive there and half the day to hike in. Stevo had, for some reason, decided to bring a six-man tent, and lasted all of 200 metres of the hike before ditching it, as my heart sank, knowing who he'd now be bunking with. We got to the campsite in the afternoon, and I walked onto the sand to find the most perfect beach I'd ever seen. I'd just seen *The Beach*, and, like Leo, I sank melodramatically to my knees at the silly, silly beauty of it all.

Tropical turquoise waves pooled themselves onto flawless white sand stretched in a perfect half-moon crescent, ringed at each end by crimson-sheened rock faces. Ridges carpeted in tea trees, eucalypt, granite boulders and endless mountain ash rose fiercely in the distance. It was immaculate and almost sacredly unspoiled, its equatorial beauty so out of place I suspected we'd wormholed into far-north Queensland. I sat on the gleaming sand and grinned dumbly until my face ached.

Stevo showed up with some biltong and a fatly rolled joint, which we smoked slowly, before swimming in the cold Southern

Ocean, contrasting deliciously with the searing heat of the summer sun and the sativa roaring through my veins. As I lay there starfished on the sand, I sighed at that slow, delicious pull of nothingness breaststroking through me; that silken thief that gobbled up time and left you with only the hazy, beautiful *now*. Slowly, the previous months of neon-lit madness and caged insanity dissipated, melted away by my endocannabinoid receptors and the hot fingers of a fierce but loving sun.

Lying on the sand, I turned to Stevo, who had just finished the last hit on our joint. His face was afire, beatific and almost saintly as he sat there, staring out toward New Zealand at a sky lit the same laughing orange as his crazy old ute.

'Y'know, when you're really high . . . it's like you're a giant penis . . . and the whole world's one big vagina.'

True to Stevo's word, the next thirty hours were a complete Gumbleton-free zone, our only companions wallabies, tiger snakes, sand flies and eels, twisting in the tidal pools by the dozen.

Something else was also true to Stevo's word or, at least, noises originating from his mouth: the sound of a snore can rise to 69 decibels, which the internet tells me is as loud as a pneumatic drill. Without some kind of measuring device, I can't tell you if Stevo topped that, but I do know the distinctive aural melange of a Harley-Davidson that won't start properly and a large animal in its violent death throes when I hear one. A feral bush pig, I think.

But even that cacophony couldn't dampen my spirits. I unzipped the tent, took my sleeping bag out onto the bone-white sand and gazed up at the milkshake smear of millions of stars, falling asleep with the wilds of Bass Strait roaring in my ears.

I returned to work refreshed, renewed with a powerful sense of optimism, and ready to be the best damn servo guy ever. It lasted about two nights.

CHAPTER 7

TAMBOURINES ON FRONT LAWNS

At the time, our store was struggling to find a permanent graveyarder to work the nights of the week I didn't. Many came and went, sometimes inside of a week – squinting flotsam on a grey tide. They generally fell into one of two main categories: disconcertingly unstable goons with dancing campfire eyes or drooping drones with as much presence as a wet leaf of lettuce. Damon was the first, a sneering rebel with dead eyes and greasy skin. He definitely looked the part: he styled his hair in an Elvis bouffant and wore a torn leather jacket when he was stocking the fridge. He liked to leave it on long after his fridge duties were over. Damon's only real problem was that, like with all new night guys, he was given a friendly CCTV viewing by our boss, just to make sure he wasn't bleeding the store dry, which he was, profusely.

He was replaced by a slim, neat man, who upon hearing the first screech of tyres, leapt a whole foot into the air then turned around to see if I was looking. He was gone in three days.

Replacing him was someone who should never be permitted to work in an organisation that deals primarily in explosively flammable materials. Joel was a pubescent character with a very *Jackass* view of the world, who didn't know the meaning of fear, but then, he didn't know the meaning of most words.

The night I met him, I approached from the carpark toward the store to start my shift when a desperate voice filled the forecourt from the amplified loudspeaker, full of tension and drama. I ran into the store, thinking something serious had happened, only to find Joel yelling rapidly into the forecourt microphone the numberplates of approaching cars like a helicopter pilot being strafed by ground fire from the Vietcong.

'Alpha Foxtrot Whisky.'

'Sierra Tango Charlie!'

'Going down, I repeat, we are going down . . . to 99.9 cents per litre, for five minutes only! That's right! Get in! Be quick! These prices won't last!'

But no, Joel wasn't on drugs, he was just a certified fruit-and-nut bar. Sort of like a young male version of Donna but with no executive function in his higher cortex. He'd make me chuckle constantly, then sigh at the looming repercussions, as he hid in the fridge to seize the hands of unsuspecting customers and laugh like The Joker as they reached in to grab a Pepsi, or when he answered our phone with, 'Footscray Abortion Clinic – No foetus will beat us!'

Joel wasn't big on the whole knowing-what-was-going-on-around-you-thing either. I came into work several times in that period, through an unlocked door, past a line of people staring angrily at an empty console. After serving them I'd find Joel in the storeroom, wandering around aimlessly or on his phone.

He was followed by Vikram. Straight-laced, sensible and very proper. A banker's haircut and an immaculately pressed shirt. He actually worked in a bank during the week. He was doing graveyards at the servo to save extra money for beginning his new life back in India, where he was to marry his arranged wife. He showed me a photo and she was a nice-looking girl. He'd lucked out, he said – as some of the partners in arranged marriages are not always as good as what you'd get on the open market, so to speak. Anyway, everything was going swimmingly for him, or so I thought, until he showed up swaying one night at 3 am with a VB longneck nestled in a paper bag and a big sozzled grin on his face, still wearing his work uniform.

'David, I am in trouble,' he said, as he dropped the bottle that smashed onto the tiles. Undeterred, he reached into the plastic bag he was holding in his other hand to start on another, and confess.

Like others I'd known, Vikram had never drunk alcohol before moving to Australia. But now he had the money and the freedom to waste it, he was like Pinocchio on Pleasure Island, getting sloshed every night he could. Instead of saving money for his married life back home, he actually owed money on two maxxed-out credit cards, mainly from continual cash advances which were then spent on St Kilda's finest sex workers. To add to his troubles, the woman he was supposed to marry was not the woman he wanted to marry, and his mother was making the trip from India the following fortnight to stay with him until he left. Somehow, she didn't kill him when she found out about his secret, ruinous hedonism, and he got on the plane and survived.

Months later, I was just about to unpack the newspapers when the phone rang.

'David speaking . . .'

'David?'

'That's me.'

'It's Vikram . . . In India.'

'Hey man! How are you doin' over there?'

'I am good. I am in Bangalore. I am drunk. My brother, he is here with me. He is drunk also.'

Anyway, Vikram ended up marrying the girl he wanted to, much to his family's displeasure. He also started up a lucrative business organising visas and accommodation for young Indian students who wanted to come and study in Australia. Last time I spoke to him he was making an absolute killing.

After Vikram came Luke. A pleasant, unassuming, tubby guy, he'd been in the navy before and was freshly shaven, disciplined and well-behaved. His mobile would go off constantly. When I asked him if it was his girlfriend, he told me his brother had stolen his phone and signed up for one of those SMS porn scams that cost $5 a message. He'd be in the middle of stocking the chips or trying to placate an angry taxi driver and his message tone would sound, promising 'tight anal action', 'norty MILFS' or the identity of his 'secret gay crush'. All at $5 a pop. He didn't know how to stop it so he was always rattle-brained. In the end, he couldn't hack the pace of the servo and ended up becoming a concreter.

Following Luke was Hazeem. He was a quiet, tall and amiable man from Quetta whose only problem was that he already had a full-time job – a fast-paced accounting job in the city – and the nightshifts were seriously affecting his daytime abilities. But because of his seven children he continued on, even as his mistakes and drive-offs increased by the day. During tax time, he was a nervous wreck.

After several more desperate randoms we finally got a steady night guy to work the graveyards I didn't. Andrzej was his name. He was Polish, ponytailed, swarthy and sullen. Hard as a cat's head. His previous job was as a pizza delivery guy, where, on what turned out to be his last night, a road collision with a woman he insists was high on crack led to him voicing such sentiments aloud, resulting in her well-built, psychotic skinhead partner silently but viciously beating him within an inch of his life.

Because of this, Andrzej was an angry young man. You could almost see the thunderheads massing above his ponytail as he stomped around the store. It became his vehicle of revenge – his projection of violent trauma onto others he felt deserved it, all from the safety of his own personal cage, not that he used it much. As soon as he got in each night, he'd stalk the aisles in a red haze, mopping the floor, even though midnight is peak hour craziness for weekend graveyards. He'd just get started and then some customer would walk apologetically through the bit he'd just done and he'd lean on the mop, eyes filled with venom as he wished death upon their children's children. He'd punctuate this by following closely behind them, mopping up the new footprints as they appeared, kicking the yellow bucket forward with his boot and flicking dirty grey water from the mophead onto their ankles, exhaling exasperatedly until they were out of the store.

Others would ask for the toilet key and, after they were refused out of pure Andrzej-flavoured spite, flippantly said they'd just go piss outside. Andrzej would stare them down and tell them they'd be doing nothing of the sort, following them outside to ensure compliance.

When actually behind the console Andrzej turned it into his personal trench; loosing verbal mortar rounds and grenades at anyone he felt the need to. He'd cut through each customer's blather with a sharply worded question, designed to shut them up and get them the hell out of the store so he could get back to his book. He brought novels into work. Liked to read the classics. The first night I saw him he was carrying *To Kill a Mockingbird*. Then Orwell's *1984*. A few weeks later, as I was putting the empty cash till into a spare drawer, I noticed a bottle of Johnny Walker Black Label sticking out of his bag.

A few months after he started, I was serving one of my last customers for the night on a cover shift when a rake-thin Fubu-hatted kid walked straight past me and out the auto-doors with an unpaid pair of sunnies, a phone charger and four bottles of Red Bull in his hands. He turned to face my shouting and cackled like a hyena then ran off into the night. Andrzej, who was coming out of the storeroom, heard my shouts and – after grabbing the metal pipe under the counter – bolted after him. Knowing he probably wouldn't catch him on foot he ran to his VN Commodore and, narrowly missing the LPG bowser, screamed off in pursuit, spending twenty minutes hunting for him, trying in vain to expunge the ghosts of his silent skinhead.

One part of the job I never considered aiming for was the position of the 2IC (second in charge). While it appealed to the Ralphs of the world, I knew it for what it was: a cleverly disguised trap. Still, all those eager little ladder climbers who felt a raging need for those numbers and letters on their name badge coveted the damn thing, giving them their first intoxicating taste of power over others. I reckon if Hitler worked in a servo, he would've started out as a 2IC.

When the position was first created, an excitement buzzed around every store as the long-time workers, especially Ralph, began eyeing each other off in competition for the prized position. The store became wrapped in a multilayered web of Machiavellian intrigue and hilarious passive aggression. Suddenly almost everyone was working hard, toiling under the watchful eye of the cameras, working extra hours and stocking shelves like a man/woman possessed. In the end the job went to Nellie, not because she had wanted or even applied for the damn thing, but simply because she was the longest serving worker on the site. Having a fair idea what it would entail, she tried to refuse it but was essentially offered the choice of 2IC or Centrelink.

As she suspected, the job was utter hell. You still had all the responsibilities of a normal servo worker, having to serve, stock, sublimate and smile, but now also had to complete a multitude of office work tasks, while ratting on and disciplining your fellow 'team members'. All for a wage increase so small it was insulting.

Every weekend, Nellie would trudge in, unpaid, trying to catch up on the work she couldn't get done during the week. She'd arrive just as I was finishing in the morning, and would stay half the day, buried in paperwork. She became snappy, terse, haggard. Nothing like the Nellie we knew and loved.

The first time I met Thomas he walked into the store announcing to all who'd listen: 'I didn't *want* to pay seventy dollars to see Disney on Ice but I just *had* to see it before it finished!' He was a weird one, Thomas. Once, when asked by an angry old man one morning why he needed to write down his numberplate, he looked him square in the eyes and said, 'So I can come and play the tambourine naked on your front lawn in the last orange light of day.' Thomas was followed by a girl who would only write with her own coloured pens in fragrant ink. Purple watermelon, it seemed, when you had disappointed her.

And then there were my immediate bosses – the store managers. I outlasted quite a few of them. While some were quiet and chilled – like my first boss, Terri – others were sticklers for rules and would glare at you if you let the lines get too long. Then there were the ones who would laugh in a high-pitched giggle then launch a steel stool across the console to make a point.

Barb definitely preferred the latter form of expression. Like so many people associated with servos, she ran on sugar, rage, dark humour and adrenaline. An uncompromising and often rambunctious woman, she was prone to stress-induced mania, leaning often on her only two obsessions – tiny Kinder Surprise toys, an army of which decorated her office, and the vicarious brutality of rugby league football – to calm her nerves. She actually tackled me once, and let me tell you, I stayed tackled for quite some time. Such a duality was the perfect window into her soul, as her gruffness and manic rage competed hourly with her undeniable warmth and thespian soul.

She had good reason for the rage, though. For, while managers don't usually serve customers, they *do* have to sift through every little bit of paperwork produced by their addled little worker bees. And if your bees are a collection of assorted misfits and sleep-deprived slaves with two and a half jobs, or seven kids, or full-time uni courses, days in the squalid hive of your office can be Damoclean at the best of times, as the demands from above filter down in their bright, passive-aggressive tones, ordering you to scold and whip your disinterested workers into achieving the impossible, while constantly looking to cover an infinite number of shifts deserted by unreliable vagrants who come and go, blowing with the wind.

Like Nellie, Barb favoured big scrawled notes in the diary – often written in hip-hop lingo – promising unemployment for the entire weekend staff, as she raged through the store like a tsunami every Monday morning. Very early on in our professional relationship, she christened me with the affectionate sobriquet 'Trouble', as even when I tried not to present her with any new amounts of it, I almost always failed dismally.

We generally got on though, as she understood us graveyarders spent hours expertly herding the unhinged and preventing catastrophe, and would usually cut me some slack. She knew when to steer clear of me, as I undoubtedly did with her. Now and again, though, she'd have to come in early, on four hours' sleep, to oversee some new equipment update and I'd be at the door to greet her at 3.30 am, shit-eating grin plastered across my smug face.

One of these mornings, as she entered the store with a bleary scowl, she marched straight to the fridge and yanked it open, grabbing a can of Coke, to hold it in front of my nose and snarl: 'Get that grin off your face or I'll shove this up you and turn you into a soda fountain.'

CHAPTER 8

AGGRO

It was a hot night in February 2004 when I realised that not only was I not in Kansas anymore, but that Toto was a snarling rottweiler on day three of the canine equivalent of an ice binge, who'd rip the straw from the Scarecrow's throat in an orgy of destruction and howl, lifting his muscled leg to piss on the scattered remains.

I was staring worriedly out the glass at a beast of a singleted man yelling at no one in particular. He had the hood of his steaming ute up and was circling it like a shark. Every few seconds he'd stop, double over in rage and scream, *'FUUUUCK!'* at the top of his lungs. People filling their cars turned their eyes to the safety of the forecourt's petrol stains, hoping that by the time they were finished he would disappear. But it never works out that way. The psycho stomped to the service bay and grabbed a wad of paper towel, fifteen sheets in his massive hand. Then, as an afterthought, he picked up one of the dirty window washers and launched it at the car wash, then hurled a full watering can at the LPG pumps, turning to face his audience, eyes burning, daring them to offer the slightest spark.

At the same time, a man who had already paid for his fuel was creeping away from the pumps in his cute little Daihatsu. He pushed the button to activate his water jets, which sprayed his windscreen, along with spritzing the muscle-clad psychopath stalking back to his ute. The hulk let out a mighty howl and chased the departing toy car. He drew level with it as it slowed for oncoming traffic then launched himself at its side mirror with a flying kick. The driver stopped in shock and – seeing the homicidal rictus grin on the psycho's face – quickly sped off, chased by a barrage of unspeakable threats, his crumpled mirror hanging by a single wire.

Servos have always been angry places. It's almost as if gas floats down the aisles, ready to ignite the sparks blown in haphazardly through the electric doors. Part of it is to do with road rage. Everyone's been stuck behind a nuff-nuff crawling 35 km/h down a one-lane street or slammed their brakes on to avoid getting jack-knifed by idiots who sail onto roundabouts without looking. And we all know that's exactly when your petrol light starts flashing at you.

And when you get to this dull glorified carport, there's usually a queue of similarly frustrated individuals honking their discontent. People hate waiting for things and even doing it sitting down, complete with music and climate control, seems to do little to blunt their anger. And then there's the price of petrol: yes, it sucks, and it's patently unfair, but you still *need* it, and going somewhere else will only you cost you time, sanity and more petrol.

Once you get out of your car, the acrid bouquet of petrol vapours steals its way up your nose. You stand uneasily, an island in a sea of shunting traffic, holding a grimy pump, with your choice of view comprising the dollars rapidly disappearing from your bank account or the unevenly hairy neck of the porcine beast suggestively scratching himself on pump four.

After trying, but ultimately failing, to not drip petrol on your car's paintwork, you brave the impatient traffic and walk toward the store, only to realise you've forgotten your wallet or purse. After returning to your car, you finally get inside and join the back of a queue that stretches halfway down the toiletries aisle, replete with uncaring mothers and their screeching kids, potential paedophiles and strangers emanating pungent and mysterious smells. The newspaper you wanted is, of course, absent.

The queue, leading to the only register open, moves so slowly you could measure it against annual coastal erosion. Your feet hurt and the store is too bright, full of shrill beeping, impatience and the people who worsen it: sifting through their seventeen discount dockets like a woman lost in the nostalgia of her old wedding photos. Almost at the front, you remember you need milk, but know the angry line that stretches behind would lynch you if you so much as attempted it.

You grit your teeth and wait several minutes while the person directly in front of you attempts to impart their meagre understanding of politics to the poor soul behind the counter, waiting patiently for the idiot to collect his change and move away. As he eventually does, the nervous but exasperated worker eyes your $50 note with concern, asking you if you have anything smaller. He then rapidly fires more questions at you about discount dockets and the pump number you forgot to look for. He scowls at you, then pesters you to buy two chocolate bars for $5, while snatching your $50 away from you and slowly, maddeningly, gathers up a mountain of coins and $5 notes as punishment, dumping them into your hands and glaring at you as you struggle to squeeze them into your wallet.

You finally turn around to be bumped into by the impatient wanker behind you, red beard like a rash, who expects a sorry but doesn't even wait for it before squeezing past you with his three 2-litre bottles of Fanta, Pizza Shapes and seemingly every bag of confectionery Allen's has ever produced. You dodge darting children to get to the doors and revving cars to get to yours, which is now trailed by several others, willing you forward with their angry eyes. You get into your car and after leaving the pump, you wait several minutes at the exit as the unwilling traffic crawls past you, eventually letting you out and you gun it, letting loose a war cry as you rejoin the race home.

Still, it's no excuse. Servos may not be the most desirable of places to frequent, but why do they pulse with such an under-current of *rage*? In how many other retail jobs are you separated from your customers by a cage, forced to wear a duress pendant at all times, and have the cops on speed dial? At times I felt like I was in a shark cage, bobbing in the ocean as the white pointers gnashed their teeth and sailed past me with dead black eyes.

I'd survived the job for a year when I really started coming face to face with rage consistently. Looking back it was, funnily enough, around the time that the price of a barrel of oil rose dramatically. As full tanks turned from $60 into $90, then over the sacred $100 mark, the sneers and choice comments grew. War was being declared by customers on the only focal point they could identify – the humble console operator.

As the teeth clenched and the anger mounted, I began to worry. Gumbletons were one thing, but so many raging nutcases released their fury only when they stalked in through my auto-doors. And it would take only the tiniest thing to light their fuse and send them screeching toward me demanding their pound of flesh.

Like a seven-foot-tall, Eastern European woman complaining to me that she didn't receive her fuel discount. She honked accusations, her square jaw set like concrete as she towered over me. I printed her a receipt and underlined her discount, but still she stayed in the store for twenty-five minutes, ignoring the waiting customers and equating my supposed mistake with that of deliberate genocide. Then there was a four-foot-tall Hobbit-like Italian man, stoned beyond all recognition, who, after discovering his pack of cigarette papers would cost 75 cents, threatened to kill me and all of my closest relatives, sliding a finger across his hairy neck and then gesturing violently at the tub of Chupa Chups he must have assumed was my head.

One woman took me not gifting her several dollars for the public phone as a personal affront to her integrity, especially considering I voiced my doubts that they would actually be used for a phone call. She returned several hours later with her own money, 50-cent coins that she pelted at my face, one after the other, to make her point.

These running battles begin to change you. When I started the job, I was a timid, peaceful, conscientious arts student who would apologise to customers every ten seconds for the mistakes I had made, the mistakes I was busy making, and the mistakes I was certain I would soon make in the future.

But as that new year bloomed into a sweaty summer amidst the pressures of countless essays, my growing lack of sleep and a house full of constant anger in the screaming death throes of a doomed marriage, I found myself pushed solidly toward a realisation that certain people didn't deserve an apology. Too many of them would burst into the store like I owed them something, ignoring my greeting and issuing their demands and any repercussions if they were not met promptly. Although it takes a while to form, the mindset of 'you can take a long walk off a short pier, you insolent twat' soon becomes something easily called upon when you need it. After a certain point, every customer I served was silently judged in regards to their distance from reaching this line. Once they crossed it, all bets and social niceties were off.

There are times when your whole being is crying out to tell the idiot how much of an insignificant gumnut they really are. The rush of adrenaline can overpower you. But in some deep forgotten fold of your prefrontal cortex, you know that the seven eager cameras and functioning microphones are watching you intently, and that your very livelihood is perpetually hanging by a thread. But then you just as quickly remember you could get a job at another servo tomorrow.

This transformation doesn't happen all at once; it's more of a slow-burn rite of passage for every young well-meaning serf who works in retail, from the pimply, ham-fisted burger flipper at Macca's, to the gangly mumbling drones that shuffle around your local Kmart. You start there and you're like a Labrador puppy: all smiles, full of beans and eager to please. But six months in, maybe a year at most, you come to a somewhat sobering conclusion: customers suck. You realise that these people actually *relish* their power over you. And as sure as night bleeds into day, they will abuse that power. It's like we're just bugs under their feet they know they can squash, so they expect us to roll around on our backs and wave our little legs in the air in the most satisfactory manner.

Not every customer is like this, granted. But so many seem to take a perverse delight in projecting their own lack of control over their tragic lives and transforming into a power-hungry douchehound, all to assuage their wailing egos. They'll gaslight, harangue and hurl insults, reducing a poor young retail worker to tears, simply because of some fangled myth that they are a King or Queen who is Always Right.

When you work in a servo, the ratio of these dickheads increases exponentially, as do their threats, rudeness and continual bleating. As I endured more and more of it on my shifts, my apologetic demeanour ossified into something more resolute. I joined a gym, started working out and taking testosterone supplements. The small amount of muscle I gained on my scrawny frame gave me more confidence, and I'd look the idiots square in the eyes, just like Siva told me to, politely explaining their options to them, brooking no nonsense. Sometimes it worked and sometimes it didn't – where high noon would last until sundown (or sun-up) – but it stopped customers walking all over me.

Nellie noticed the change in me first. Her eyes raised from her *New Idea* mag and her slices of dragonfruit one night as I politely but firmly dealt with a whingeing, entitled P-plater, and again I felt her measured stare.

'You're different, David.'

'How so?'

'You're not putting up with their shit anymore.'

'I don't think I've had much choice. This place is kill or be killed.'

'Yes. I think that describes it perfectly. Do you remember I told you that you would change? And maybe not for the better?'

My assertiveness soon spread itself to other areas of my life. I was always a shy and withdrawn kid who'd avoid conflict at all cost, even if it meant I got screwed over in the process, and this had always bothered me. But after a year of being left alone in the shark tank to sink or swim, I found I was developing several new strokes with ease. Or that I'd finally found my speargun.

Compared to my freaks and psychos, the general day-faring public was nothing. Face up to most people behaving badly and they'll turn away with their tail between their legs.

Sometimes, I'd be waiting in line at a supermarket, a department store or even another servo, and some cretin would be harassing the poor retail worker, whingeing about a price increase or the fact that their arsehole was only tongued six times instead of the usual nine, strafing the poor check-out chick with gleeful threats of unemployment. Nothing made me angrier. What kind of scum-sucking vermin goes around seeking egoic validation of their brattish inner child in the process of threatening to get people in shitty, low-paying jobs fired? I'd butt in and explain, again politely, but also quite firmly, that yelling at the poor kid was not going to get them what they wanted and that buying frozen chicken tenders and a discounted pack of double-coated Tim Tams did not grant them a royal decree to act like Piers Morgan on an ether binge. As soon as I intervened, I saw the look of relief on the worker's face as they realised this idiot was no longer their problem. We would both watch them stalk out, threatening legal action or more violence, as we grinned and high-fived in solidarity. Sometimes we even hugged. It was nothing ground-breaking, of course, but for me it was indescribably huge.

Although basking in my newfound confidence, I was at the same time a little worried about what I was becoming. Nellie was right, of course. Each running battle would leave me appearing forceful and in-control in that moment, but due to the cyclical nature of customers, they'd return for round two, then three, ad infinitum, and soon I was heaving with the draining effect of all this perpetual conflict. For a peaceful arts student, this rapid transformation, while empowering, wasn't really enjoyable at all.

As it began to build, I asked Rathindra – the Bangladeshi Mr Happy – how it was he kept his peace, especially in the midst of such unrelenting belligerence. He simply smiled (well, more than he was already) and stopped counting his till as he leaned back wistfully against the cigarette cabinet.

'Well, David, I suppose that I have learnt to be thankful for whatever life throws at me. She is capricious. She can be brutal, unfair, yes? But she is also very generous if we can be grateful for what is given. I like to live in the moment that is given to me and take all I can from it, because *this*,' he waved his hand across the desert of seething colour, 'is where I am. Right here, and *right now*. It is *always* now.'

'Yeah, but right *now*, wouldn't you rather be somewhere else? I mean, this servo and all these idiots can't exactly be the best company in the world?'

'I suppose I think of the quotation that says to "be kind, for everyone you meet is fighting a hard battle". I know nothing of these people's problems, their pain, their sadness and regrets. And in addition to this, I enjoy learning from everyone I can. Everyone has a lesson for someone. We just need to open our eyes, our minds and, I think, our hearts, so that we can learn it for ourselves.'

'And how do I do that?'

'Well, I like to try and still my mind.'

'Mmm, that sounds nice. How do you do that, though?'

'I practise yoga, which is useful, but mostly I meditate. I find that it calms me and clears my perception so that my ego, which depends on emotions like anger, isn't running the show.'

'Well, it seems to work smashingly for you.'

'Really?' he said, beaming at me. 'Well, yes. I suppose it does.' He then turned to a shaven-headed truckie, who returned his smiled greeting with a sedated scowl.

While Rathindra gained his peace from meditation, I seemed to find only more anger. I started forming views about customers that maybe, looking back, weren't particularly helpful or constructive. Such views were encouraged by Nellie, smiling craftily at me when I was ready to lose it with some arrogant fool, before recounting her own stories from years gone by as I burnt holes in the back of their departing shirt with my eyes.

Nellie was almost strangled by a psycho once. He walked in, grabbed an ice cream and walked out the door eating it. When he came back in, Nellie told him – politely – that he really should pay for his Magnum first, and he told her to shut up and mind her own business. He paid for his diesel with a fuel card and then – ignoring the many notes in his wallet – got his credit card out, but it was declined. He smirked and said he couldn't pay for the ice cream after all. Shrugging, he turned away from the counter and headed for the door, taking another goading bite as he did. Nellie locked the door, smiled and waited, as he walked into the glass with a rattle. He marched back, spittle and ice cream flying as he demanded to be freed. Nellie told him he'd first have to pay. He threatened her, and when that produced no effect, he pelted the chocolatey-white lump at her head before jumping the counter and knocking them both back into the cigarette cabinet, his huge hands around her tiny neck. Nellie kneed him in the balls, ran into the fridge and called the cops. This sort of thing is why servos now have anti-jump wire.

Another time, she was waiting to serve a man who'd walked into the store on his phone, his rottweiler trailing behind him. As Nellie told him his dog wasn't allowed in the store, he waved his hands dismissively, saying, 'In a minute, love, don't get ya fucken knickers in a knot.' Nellie repeated her demand and was again ignored. She did it a third time, her screeching Serbian voice shattering the man's reverie. He forgot about the call and walked up to Nellie and spat at her, the glob landing on her cheek. She just looked back at him and did nothing, meeting his stare but staying perfectly still.

'I didn't want to wipe it, David. I didn't want to give him the satisfaction, so I just stood there and felt it dripping down my cheek and onto my chin as I stared at him.'

Such tales only added to my rage, and I felt myself rapidly turning into Andrzej, my ego itching for a fight with anyone who walked through the auto-doors with or without something to prove. I started speaking in short, clipped tones and would only assist those deserving of it.

Trying to forget my troubles, I headed out with Stevo to the footy. We were at the MCG, the vast seagull coliseum that, on this miserable winter's day, was three-quarters empty. The Western Bulldogs were in a dark place, having declined in form each consecutive year to the present nadir and, while one flag in five decades meant watching them play was ultimately an exercise in self-flagellation, today it bit down harder than usual.

A cold wind whipped around the gargantuan stands as seagulls hunted chips relentlessly, squawking in need. Goal after goal sailed over our heads as we downed beer after beer, watching anyone in red, white and blue being pummelled into submission by a Melbourne Demons team who were nothing special themselves, but made to look like Harlem Globetrotters by our bumbling ineptitude.

Stevo gestured to one of our young forwards lining up for goal, who duly missed. 'He looks like the kind of guy that still collects Pokémon cards,' he said, earning a few sniggers from Dees and long-suffering Bulldog fans alike. With a chain of handballs and one long kick, the Dees went straight down the other end and added another goal to their growing score.

'Dude, this is brown,' Stevo moaned. 'We're being reamed. If I wanted to see sexual assault I woulda just gone to King Street. Imma go snap one off.'

He fished into his thick sock and pulled out his hip flask.

'You need this more than me.'

I watched on as, straight from the bounce of the ball, an insouciant Adam Yze grabbed the footy and slalomed through packs with consummate ease, brushing off our pathetic tackles like dead flies from a windowsill. I cursed, sipping Stevo's flask of terrible sweet bourbon and wondering how bad we could get. All around me incensed cries rang out from Bulldog supporters fed up with the insipid display. Tired of encouragement, they began to switch to mockery and rage.

'You're pathetic, Darcy!'

'You couldn't get a kick in a martial arts film, Brown!'

'Grow a pair, you spineless frauds!'

'See those big sticks, Eagle? Ball goes through there, you fucken clown.'

I kept downing the bourbon and soon I was laughing to myself and joining in. I watched these irate supporters, leaping from their seats as the footy trundled over the boundary line, shaking fingers stabbed accusingly at the culprits, their spittle flying and eyes kindled with rage.

I thought then about Rome under Augustus: gladiators hacking each other to bits surrounded by lions and tigers, a populace enthralled, happy and fat with the government providing them with endless bread and circus.

Staring at the swathes of irate fans – pouring out their quotas of weekly rage at our hopeless players and besieged umpires – I suddenly saw my customers. The metaphor strengthened by the second. Both groups erupted with hatred at impersonal avatars, and the whole process seemed incredibly cathartic for them. If footy was a kind of ghetto therapy for average blokes, then what was my servo for the unmoored denizens of the night? I took a swig of bourbon and laughed bitterly. Of course. What else was a nocturnal servo but bread and circus? Like Augustus, I fed my freaks free carbohydrates to keep them in line, but in truth, I wasn't even a lame-duck emperor. No, I was more of an umpire: a vital cog in the machine, yes, but universally despised all the same. No one respected half-hearted moderation and passive-aggressive attempts at control.

But I did have a whistle. And I planned on using it.

CHAPTER 9

BULLS ON PARADE

As this anger within me raged on, I found there was another group of customers that seemed to embody this more than anyone: taxi drivers.

In a way I felt a kinship with them. For we both endured the Sisyphean punishment of dealing repeatedly with the barely animate pond scum of society. While these days, Uber and co have smothered them into irrelevance, back then they were ubiquitous slaves of the night, like the nomads of old, forever journeying between city and suburbia, straining to hear news from their home country on BBC World over the raucous laughter and racist insults of their drunken passengers. All the violence and vomit they carted from suburbs to city and back again.

But familiarity does indeed breed contempt, turning most of them into sullen, selfish creatures that hardly ever spoke – no matter how much I tried to engage them in conversation – unless of course they wanted something. Their patience levels were thinner than a strand of recombinant DNA, completely unaware of the delicious irony that all of them seemed to operate on some sort of weird time-delay setting.

Almost every taxi driver who pulled in at night would take five minutes before even getting out of their car, only to trundle to the middle of the forecourt, oblivious of approaching traffic, and slowly stretch, like a hippo on Valium. After another few minutes they approached the LPG pump, often realising they'd pulled their taxi up on the wrong side of the bowser.

After pumping their gas, they returned to their car and sat inside for over ten minutes, fumbling around for the most inconvenient amount of change they could give me. Meanwhile, another taxi pulled up alongside, who appeared to be the first driver's long-lost cousin. They embraced, then spent fifteen minutes reminiscing while I ground my teeth and stormed off to the fridge.

The second I got in there, they rushed toward the store and banged impatiently on the glass. As I emerged and headed back to the console to unlock the door, they barged through the auto-doors, scornful and truculent, approaching the console and demanding 'receipt', as they dumped over $50 of change on the counter, then pushed each 20-cent piece across with a slow, measured deliberation, oblivious to the line that had grown behind them. They mumbled angrily into Bluetooth headsets while seeming to stare far away, possibly at somewhere they'd left to come here, wondering if the grass really was greener.

After the mountain of silver had been counted, they snatched the receipt out of my hands and perused it through their small glasses, assuming the servo was scheming to rob them of several cents a litre. Half-satisfied, they turned abruptly, usually stumbling into a waiting customer, who they grunted at, before waddling out of the store.

One cabbie, Javad, four-foot-tall with oversized Coke-bottle glasses and a majestic sense of self-importance, possessed the preternatural timing of always showing up when my mood was so toxic that I felt like eviscerating the small and fey and dancing in the scarlet spray. And he would always push things. Just that little bit more.

The first time I met him, a steamy February morning in 2004, he marched into the store and demanded I fix his phone. I looked down at him – his black eyes magnified through his thick glasses, searching me for a challenge – and realised that he was serious. I tried to explain to him that that wasn't my job, and even if it was, the inner mechanics of Nokias were well above my intellectual pay grade. It was at this point that his supposed broken phone rang, so he answered it, ignoring me as he yelled into his Bluetooth headset in Punjabi. Finishing his call, he looked me up and down, declaring, 'You no good', before walking off in a huff, back to his cab. He returned a minute later waving a $100 note, yelling, 'You change! *Now!*' The next weekend he came back, this time striding through the opening glass doors with a verbal list of things I was to fetch for him from the shelves. This went on for several months.

There was also this young driver that started showing up. Every Friday and Saturday night, he would pop in, dressed up like he was about to go clubbing: Issey Miyake aftershave, silky shirt and tight-fitted slacks, hair waxed into a cavalcade of perfectly messy spikes. His taxi, though, was even cooler than he was. Lowered, it had a spoiler and what looked like 19-inch rims. Its quality sound-system poured R&B anthems beseechingly out into the night. The poor guy was just itching to get out there among it, but had to spend all his Friday and Saturday nights copping crap from drunken Aussie yobbos.

Even before the drunken yobbos got into a taxi, they were already playing games with the poor souls trapped in the driver's seat. They'd stagger into the store after pre-drinking at home all night, inhale several cheeseburgers, then get me to call them a cab. Before the days of Uber, when you actually called for a taxi, you had to supply the company with a passenger name. The drunks, of course, saw this as their creative outlet, giving me the most ostentatious and downright bombastic names they could muster. I'd end up saying to the jaded and impatient woman at the cab company: 'Can I get a taxi to the city for . . . umm . . . Cornelius . . . the Second.'

'Has he got a mate with him?'

'Yeah . . .'

'And what's their name?'

'. . . Captain Moon Star.'

(Audible annoyed sigh) 'Any others?'

'Umm, yeah . . . Smithy . . . the Longshanks.'

One morning that summer, as a Saturday night was slowly becoming a stinking hot Sunday, a taxi came screeching onto the forecourt, locked tyres smoking, narrowly missing another driver who was filling up. Screams and rage-filled threats poured out of the cab, followed by a drunken man and two pregnant, also drunken, women. They'd presumably been fighting with the cabbie over the fare or God knows what. By this stage he'd obviously had enough.

Once out of the car, the pregnant women started taunting the driver in unison, hurling racist epithets while casting aspersions over his driving ability, the more eloquent among them telling him he couldn't drive a greasy stick up a dog's arse. They continued on like this for minutes – as the customers on the forecourt busied themselves with their bowser readings and their petrol caps – trying to entice him out so the man could get a decent punch in. One of the women then came running into the store, announcing that the taxi driver had ripped them off and, bourbon swimming from her every pore, ordered me to call the police. A minute later the other one burst in shaking her fist and slurring that if I've called the cops then I'm gonna get what's coming to me.

By this stage, the man had thrown a few quick jabs through the open window before the cabbie could close it. As the cabbie rolled it up and went to drive away, the dickhead jumped in front of the cab, blocking it, hurling vile racist insults and taunting the driver to be a man and run him over.

So he punched the accelerator, loosing a fierce war cry as he did.

He could have easily reversed and driven around him, but one thing you learn about nights is that everyone has a breaking point and cabbies are pushed further than most.

He hit him hard, flipping him up into the air like a rag doll. Not that this deterred the guy. He got straight back up and told the cabbie to do it again. So he did.

This time the guy got up slower, limping now, while his two pregnant companions had started arguing with each other, presumably about why one wanted to call the cops and the other didn't. They started pushing each other, before graduating rapidly to trading roundhouses. Neither of them had noticed that their male friend (possibly the father of both of their children, such were his evident superhuman powers) had been mashed twice by a car and was limping toward the vacuum parking bays.

At this point I figured I should probably call the police. After I'd informed the triple-zero lady of my emergency and my address, she mentioned that I sounded rather nonplussed about the whole thing. I suppose it happens after a while. As usual, the bogans had disappeared long before the cops showed up, hopping into another unsuspecting taxi.

So, with my newfound balls, I was beginning to use my whistle adroitly and assert some order and ownership over my aisles. Ten minutes was the maximum non-regulars were allowed to stay. Good behaviour was rewarded with a free pastry, slurpee, or even a rare chocolate bar, if it was earnt; a true Gumbleton Happy Meal, while anything out of line was stamped out before it had a chance to spread. While there were many disagreements at first, my clientele eventually began to respect me, acquiescing to my demands after I had, in a stern but fair voice, repeated myself several times. I felt good. For eight hours a night, I was something loosely approximating the beginnings of a man.

But I quickly found I could only control what happened inside. The forecourt, as previous incidents had demonstrated, was a no-man's land I was happy to cede to others, where the rebels and no-hopers would lower their antlers and charge each other repeatedly.

Cars, anger and testosterone. They go together better than toast, butter and Vegemite. The acrid smell of scorched rubber from constant burnouts would hang in the air like a pheromone that attracted only hoons and psychos. I swear that whenever it rains, the smell of the wet earth flips a toggle switch in hoons' brains that lights up their neurons like a Vegas casino:

BURNOUTS! BURNOUTS! BURNOUTS!

VN Commodores would drift and slide round our roundabout – for twenty-five minutes straight. Some of them managed to stay sideways for several entire rotations, holding up growing lines of traffic from all four directions that was too terrified to enter. Of the more than 1000 suburbs in Victoria, Werribee and surrounds were near the very top of the list every time the figures were released for cars impounded under the new anti-hoon laws – fifty per week, at one stage.

They never learned. The cops would fill up several times a night, and I lost count of the times I saw a constable leap into his car, sirens screaming and his pump lying forgotten on the ground, flashing into the dark in ocular orgies of blood and blue.

But the hoons just kept on screaming, sliding, burning, drifting and smashing their way into trouble. Even when a cop was filling up, they'd sit in their cars and rev them – wasting more energy than Ricky Martin's girlfriend – ensuring the cop would simply follow them and pull them up for some obscure violation of the road rules. The cops considered my servo a goldmine in completing their nightly quota of tickets and arrests.

The servo was a saloon in those days. The heat baked into the concrete during the day would rise with the petrol fumes, as the bulls went on parade past the flickering pump lights. There was always a glut of muscle-bound, tight-topped, insecure guys in their hotted-up angry-looking cars, screeching to a halt outside my door. They'd spring from their cars (enclosing a meek-looking female in the passenger seat who was trained not to look at other males) and in through the auto-doors with dark looks on their stubbly faces, scanning the store for a challenge. They approached the register smelling like they'd been swimming in two-dollar aftershave, with a hard stare and clipped tones, as if speaking at all was an inconvenience I was solely responsible for.

'Winfield.'

'Which ones, mate?'

Pause. Stare. 'Blue.'

'Er, which size?'

Another stare, eyes filled with challenge . . . 'Twenty-fives.'

They dropped their money onto the furthest corner of the counter – ensuring I had to strain through the anti-jump wires to reach it – then snatched their smokes away, tearing at the plastic wrapping and letting it float to the ground. They then strode out with a practised deliberation, bouncing angrily on their heels to rejoin the cacophony on 20-inch wheels as they blasted out into the hot night.

I soon learned that, behind the wheel or behind the counter, everyone was fighting some battle or another. I even felt a pang of sympathy when Javad sped in one night, desperate to disgorge his drunken dickhead.

As the canary-yellow station wagon skidded to a stop outside the store, I noticed Javad's shoulders were being used as a footrest by the bogan behind him. He freed himself and implored me to help him, so I suggested we ring the police, to which he, being a militant little man, refused point-blank. He instead made off with one of the fire extinguishers and tried to bludgeon his passenger with it. When this didn't work, he tried to drag him out by his feet, copping a boot in the face for his troubles.

After a minute of this he sighed, sat down cross-legged on the forecourt with his broken glasses, and waited, staring off sadly into the night as the passenger taunted him from his own cab. Javad called another larger driver, who showed up and hauled the drunk out onto the forecourt, standing over him with an aluminium baseball bat until he staggered away in horror. Meanwhile, Javad and his mate had lost a fare each, which, when you earn two-tenths of fuck-all an hour, was definitely a noticeable chunk.

Just like cabbies, constantly hounded by both customers and their company, people working in servos often snap. While it's usually only an angry page-long note in the store diary, sometimes things go further. Sometimes, the console operator will get that animal eye-shine about them; that homicidal glint in their eyes that says: *All bets are off now, Jimmy.*

Sometimes, they're pushed and pushed by a smartarse or angry customer and they'll smile unsteadily while wishing them a pleasant night as they walk out the auto-doors. As soon as they're gone, they'll lose it completely, hurling staplers and sticky-tape dispensers at anything that moves while they howl like a werewolf, inflicting their rage on inanimate objects that won't fight back or treat them like dirt.

Speaking of dirt, as we saw with Andrzej, one source of constant warring between customers and console operators is the floor. When it's quiet, you're always supposed to try and mop, but deep down you know you'll get interrupted, usually on a rainy night by someone who pretends not to see the folding yellow sign and waltzes straight through the middle of the bit you've just finished.

Early one night, the customers had momentarily disappeared, so I grabbed the ancient yellow bucket and mop and got to work. I'd almost finished when a Daewoo pulled up and a friendly looking woman with an eighties perm, noting the wet floor sign and the poorly disguised scowl on my face, restored my faith in humanity.

'Don't worry, love, I won't trample all over your nice clean floor,' she said. 'My husband does it at home and it drives me up the goddamn wall!'

She then tiptoed to the hot-food section and grabbed several napkins that she used to deftly wipe away her barely noticeable toe prints as soon as they appeared (despite my urgings not to worry). Pleased with her good deed for the day, she approached the register to pay for her bread, milk and impulse Chunky Kit Kat purchase, but got only halfway before her attention turned to the opening doors. She screeched out with venom and spite at the newcomer:

'WHAT THE *HELL* DO YOU THINK YOU'RE DOING DRAGGING IN ALL YOUR MUD AND SHIT OVER THIS YOUNG MAN'S *CLEAN* FLOOR?!? I GET ENOUGH OF YOU DOING IT AT HOME, YOU USELESS *FUCKBUCKET!*'

The poor husband, eyes wide as doorknobs, froze to the spot and then retreated to the Daewoo, slinking away from the faint tracks of mud like a scolded child.

'I'm *so* sorry about that, darl . . . From now on, *he* can wash the fucking floor.'

CHAPTER 10

SMILE (WITH TEETH)

*In the end the servo would announce that two and
two made five and you would have to believe it. It was
inevitable that they should make that claim sooner or
later: the logic of their position demanded it. Not merely
the validity of experience, but the very existence of
external reality, was tacitly denied by their philosophy.
The heresy of heresies was common sense. And what was
terrifying was not that they would kill you for thinking
otherwise, but that they might be right.*

George Orwell, *1984*

(Well, obviously George didn't write 'servo', but you get the point.)

Anyone who's ever worked at the coalface of a massive retail
corporation knows they keep their employees in check. The
unceasingly tyrannical machine is well-oiled, the human
cogs perpetually surveilled by its unremitting array of digital

panopticons. Its increasingly ridiculous demands from on-high, cloaked in confoundingly bright sunny tones, can often seem Big Brother–like with the implacable absurdity of it all.

I was almost a year into my job and, with my newfound confidence gained from staring down psycho meatheads in the lonely hours of the night, I was beginning to question certain things around me. After an initial grace period, where I was so happy to get – and somehow keep – my job, I began to wonder if my company really did value me as an employee.

Having studied the bureaucratic wiles of Communist regimes at high school and uni, and now being deep in the fluorescent guts of Shiny Happy Capitalism, I started to see that I was mired inextricably in an unholy marriage of the worst parts of both.

What I discovered fairly rapidly was that the overlords in corporate liked to talk in *doublespeak*. That is, on the one hand, they informed us we were essential 'team members', an integral part of the organisation. Valued, trusted and respected. The whole warm, fuzzy 'happy family' angle.

On the other hand, their true aim was only to increase competition and distrust among our fellow comrades to increase our monthly quota size. We were not team members, but slaves.

Servos have always been salt mines, the workers themselves possessing no power or value. This is because they perform a job that, in terms of pure mechanics, can be done by literally anyone. In *Simpsons*-speak, we are all drones of Sector-7G, and somewhere, Monty Burns is watching. The essential part of the job is smiling, suppressing your rage into a sing-song voice, and pressing buttons. Smile. Press. Suppress. Smile. Press. Suppress. Suppress rage. Suppress any identifiable human urge. Suppress clambering existential dread. Smile. Press. *Smile*. Such workers are easily replaced, and the company is adept at first pushing them to their absolute limit, wringing labour from them like water from a grimy dishcloth.

There were constant Ministry of Truth memos spewing from the overworked printer and tacked up on the wall, saying one thing but meaning entirely another. If older than two months, they were destroyed, thus you could never quite remember what any newly introduced policy actually used to be. These memos contained reminders, scoldings, lambastings, five-year plans and brightly coloured graphs to demonstrate the company's perpetual disappointment in all of us. The thing is, though, it's pretty hard to follow a memo reminding you to 'enquire if the customer will be using a discount docket today' when the 'customer' is a psychotic deviant with hair like a ten-year-old mop, sliding into your store on a skateboard and sailing straight into the ice cream freezer, where he lands in a heap, before enquiring of you: 'Hey buddy, *WHERE THE FUCK ARE THE SAUSAGE ROLLS?*'

But in the official tomes of the company, all clean curvy lines, bold colours and smiling, supplicant drones, such creatures were *wrongthink*. They simply did not exist, thus all untenable orders must be followed to the letter, piling both the cognitive dissonance and the nonsensical work required by the tired slave ever higher until they simply acquiesced.

As I began to posit such a theory, I started to see more and more exploitation around me. It was everywhere, not only in Barb's 'don't push it' face as she blustered in on a Sunday morning to finish paperwork that didn't get completed during the week, but in the hunched posture and blank thousand-yard stare of Ralph, who had been fighting an eternal line of customers for six hours because his cover shift didn't show up and the company didn't even have the decency to arrange for someone else. They saved on hours that way, you see.

In fact, they saved on hours a lot, my company. The problem wasn't that there wasn't enough work to go around; the problem was that there weren't enough *workers* to go around. By the end of my first year of glorious service to Mother Servo, my mobile would ring constantly, a retail area manager on the line wanting to know if I'd be free tonight to work a shift in Sunshine, Deer Park or Broadmeadows; pretty much any servo where, I can confidently declare, that if time somehow escaped its linear constraints, spiralled into infinity and then turned around on itself to touch the past, imprisoning me once again in those factories of creeping dread and twirling string bags, then I would happily hammer a rusty nail into my eye then slowly eat my own face.

Not long after they start working nights in a servo, workers often began to notice the absence of something they may have become somewhat attached to in their previous position of employment: breaks.

When you work for eight hours, you usually get to have a break for lunch/dinner/breakfast. Some sort of meal. Everyone in the office, warehouse, factory or department store will, at some point, look at their watch with a sigh of relief as they head off to a mundane but alleviating half-hour of reprieve.

But when you work in a servo, you're not even allowed off site. And your 'break', if you're incredibly fortunate, is maybe fifteen minutes hunched over a bench, trying to eat a seven-hour-old pie or whatever you've brought from home, usually in plain view and earshot of the unwashed hordes, so that you can still be kept company by their unique charm as you cough down your food before getting back on the conveyor belt. Almost without fail, though, your break is cut short as a price change, a swapping of a gas bottle or some other mundane task requires your assistance, all so the other worker can keep repelling the implacable stream of grumpy, impatient flesh.

In over five years of working solitary nightshifts, I was allowed two official breaks. One of those was when Rathindra somehow locked himself out of the store just before midnight, and after two hours of waiting in the cold I went off in a huff to Macca's. The other time was when there was a total blackout and the store's back-up power supply had run out. Beyond those two situations, I can't recall a time where I wasn't at the beck and call of the night-faring population of Werribee and surrounds.

As I slowly became competent and stopped fearing my imminent termination, this really began to get to me. I'd be sitting down with a sausage roll and *TIME* magazine, or I'd be on the crapper, having just got down to business, when the glass doors would rattle and shake from hammering fists, punctuated by that droning, maddening 'Hel-lo?'. I'd leap up, cursing, and rush to unlock the door again, only to be drunkenly lectured for a full minute on why I was so bad at my job. This happened several times every night for the overwhelming majority of every graveyard shift I ever worked.

It was the middle of the night, but there was never any peace. From a grateful, eager-to-please fool, I'd grown bitter and resentful of my new masters. In fact, I can pinpoint the very moment I began to turn against them completely.

See, in my six long years of servo-tude, I called in sick twice. I'm quite proud of that fact. The first time I was in shock from a car accident. It actually took me fifteen minutes to convince them I was serious.

The second time, I had impressively explosive gastro. Stevo would've been proud. Five kilos down, I was so dehydrated and woozy that I had trouble sitting up straight. I called the area manager from the hospital emergency room and explained that I obviously wouldn't be able to make it in tonight, even putting the nurse on the phone to corroborate. The bastard sighed angrily before informing me that it was my responsibility to cover my own shift. I repeated as clearly as I could my current predicament, but he was unmoved and hung up.

After that night I realised it was just easier to show up, no matter what condition I was in. I knew the company would say there was no cover available anyway, and even if they miraculously found someone, it'd take hours for them to get there.

I ended up working shifts where after I finished and made it home, I actually thanked whatever force was in power of our vast cosmos that I'd survived and it was over.

I woke up one night to realise that my day at the footy (the Dogs lost, again), combined with some fast-acting viral infection, had left me with no voice. I went to ring the night area manager, but with no voice, this was a rather fruitless endeavour. After texting to and fro, once again I was told that my shift was my responsibility. So, I showed up to relieve the guy finishing afternoon shift and, after much gesturing and sign language, he ascertained my problem. The trouble was, he really had to go to his other job, which he'd been warned he'd get fired from if he showed up late again.

After he left, I made up a large festive sign proclaiming my muteness, then spent eight long hours gesturing and smiling at the same inane comments from customers while fantasising about attacking any manager from corporate with a witch's hat, pummelling them with it until they sang a song of pain like a crippled canary.

A few weeks later, on the night of my birthday, I started drinking a bottle of Kahlúa at 10 pm, polishing it off just as I was getting ready for work. My jubilant mood soured as I realised I couldn't drive, and by the time I arrived at work – courtesy of a taxi, driven, of course, by a stern, disapproving Javad – I was reassessing my life decisions with a newfound sagacity. I somehow got through that night, made several sacrifices to the halogen gods, and afterward decided I was clearly invincible.

I wasn't, though. Uni and work were starting to disagree with each other more than a blue-haired Democrat and a red-capped Midwesterner. I had eleven assignments, essays and feature articles due at the time and was struggling to write one every two nights at work, in between spiky-haired twats stealing slabs of Coke and wannabe gangsters brawling on the forecourt.

I'd finish work, drive home and shower and, just about to dress for uni, I'd realise I had two essays due on that very day, so I'd stay home, on no sleep, and somehow knock off pallid facsimiles of what they could have been with a functional, rested cerebral cortex.

But my grey matter was fast slipping under oceans of madness. Lecturers wore a look of surprise as I stumbled in to my first lecture in weeks, bleary-eyed and broken. High distinctions become distinctions, then credits, then 'please see me after class's.

One teacher actually asked if I was having any trouble at home, which was nice of her, I suppose. I remember replying, 'No, no, I'm not having any trouble . . . at home.'

You simply can't write an essay on capitalist intervention in the Korean War when every sentence is scuppered by sirens, squealing tyres and foul-smelling pan-handlers trying to trade their wares for copious packets of Winnie Blues.

There were others at uni who noticed the change in me. I had a friend in African American History class. Paolo was his name. We had hit it off the first day we met and shared a love of bad gangsta rap and Thai green curries. He had this piercing perception, so uncanny that I found myself wondering if he'd emptied my skull with a spectral ice cream scoop while I was sleeping and eaten my dreams for himself. He had begun to notice the change in me as my company wrung me like a wet sock.

Feeling like shit-on-toast, I walked late into class in the middle of a discussion about Malcolm X, trying to seat myself quietly as our lecturer eyeballed me indignantly. After almost tripping over my bag, I fell into my seat as the conversation started up again. As the lecturer droned on, Paolo leaned over and said softly, 'My friend, you look like a man that is addicted to his own kryptonite.' It was so trenchant I laughed aloud, earning another glare from the lecturer. I think it was the first time I'd laughed in about a fortnight.

But I stopped pretty quickly when I realised I was starting work in about six hours that night at Laverton North, a hellish store in Melbourne's outer-west industrial estate graveyard, with a drive-off every ten minutes and an endless conveyer belt of bitter truck drivers coming down off too much speed, all on next to no sleep.

I wasn't the only masochist, though. Maybe it's Stockholm syndrome, but something about working in a servo unmistakably hardens you into attempting painful feats of strength. All the rudeness, noise and spluttering threats bounce off you eventually, deflected by the shell you start growing, your rage transmogrifying into a twisted sense of pride that you can endure more stoic punishment than any other sort of retail worker and still largely function as a human being.

Andrzej (the ponytailed psycho with a predilection for scotch and classic literature) was like this even before servos, but servo work built a new kind of super-shell over the top; a self-nourishing organic coating that fed his natural warrior state and allowed him to go into battle against the hordes with impenetrable armour for hours on end.

He liked to test himself as often as possible. He once worked a five-hour shift during the day, then an eight-hour nightshift that night, then nine hours the following day, trudging into my store just before midnight on his way home to check his hours for the following week, looking like some decaying organism that had been left forgotten in the vegetable crisper of time.

He headed to the back room and saw that, due to a rostering mistake, he was rostered on for that night, starting at midnight. He rang our new area manager, Nadine, and pointed out the mistake to her, but unaware he had worked all day, she just asked if he wouldn't mind working tonight as they simply had no one else to cover. Maybe he wanted the money. Maybe it was just the masochistic challenge. Maybe he had a secret dungeon where he flagellated himself with barbed whips to sanctify his suffering soul. Whatever it was, he said yes.

After mainlining Red Bull for his entire shift, desperate to recover, he asked Nadine if he might get a night or two off, as he was kind of a bit worn out. Nadine fired right back, questioning his commitment to the company, even doubting if he actually wanted work at all. She kept at him for ages. Andrzej hung up in shock, learning a valuable lesson in what was to come from our new prison warden.

Around this time, while working by herself at a neighbouring store on a Sunday evening, Donna felt decidedly unwell. When she had a minute, she rang the weekend supervisor, saying that she might need cover to go home early, as she was feeling weak and dizzy. The supervisor said that he had no one to cover her shift, so she would have to stay until the end. Ten minutes later Donna could no longer stand up, and fell down onto the mat, hiding her from view of the waiting customers. Wondering what was going on, one regular with retail experience came around and started serving the customers herself, with Donna barking orders from a chair. Donna went to the doctors on the way home. Turns out she'd had a heart attack while serving customers and was lucky to survive.

You'd think with our dedication to the job, there would have been some kind of loyalty shown toward us. But when ten years had passed and Mother Servo decided to reward Nellie for her loyalty, what was she given for all those years of smiling servitude? A $30 gift voucher. And when she went to use it, it was declined. Apparently, there was no money on it.

To further break our spirits, we were denied anything that may have brought pleasure. When I started the job, we were allowed music. We'd happily bop about to the radio as we served customers, but the company got rid of that, citing some nonsensical excuse about security. I came in one night after a long day at uni trying to catch up, grabbed a coffee and walked to the radio to turn up Triple J's hip-hop show and wake myself up. Staring at the empty space, I turned around, distraught. I looked at Ralph, who only nodded sadly before turning back to his customers.

Soon the fight left in everyone started to fade. In the words of Orwell himself: 'In the face of pain, there are no heroes.' The light that used to linger in my comrades' eyes, memories of how things used to be, all evanesced into a subjugated nothing.

I probably seem melodramatic. But most people still think working in a servo is a breeze. I'd say it's closer to a hurricane. The following is what each and every team member was supposed to do for each and every customer, which at a busy site could reach over 1000 a shift:

- Pretend you like them and that their very presence has brightened your day.
- 'Hello sir/madam/mate etc, how are you today?'
- Pretend you care about their answer, and then answer their equally empty, inane questions with fervour.
- Smile (with teeth).
- 'Do you have fuel or gas today?'
- 'And what pump number was that?' (Customer has no idea.)
- 'Will you be using a fuel discount docket today?'
- 'Would you, by any chance, be interested in purchasing two of these for three dollars? It's a good deal after all.' (There were over eighty deals every month and you were supposed to memorise them all, along with the five-digit code that accompanied them.)

- (To customers waiting in the line): 'Sorry, won't be too long.'
- Swipe customer's card.
- Request them to enter their account details, their pin, if required, and to press OK. (End up having to do it yourself because customer is a braying fool.)
- Check signature against credit card.
- Check credit card against a list of stolen credit cards.
- 'Will you require a receipt with your purchase for tax purposes?'
- 'Thanks for that.'
- 'Have a nice day, see you later.' (Sincere farewell.)
- Smile (with teeth).
- (To next customer): 'Sorry about the wait . . .'

And as the next lemming approached, it began anew.

Although before the customer even entered the store, you were expected to:

- Observe and record the numberplate of the vehicle.
- Observe and record details of the car (i.e. colour, make, model, sedan/wagon) and its occupants.
- Observe details of driver (clothes, hair colour, height, etc).
- Check numberplate off against list of stolen plates.

On top of this, after eight o'clock every night you must also observe every single customer who approaches the locked door and scrutinise them individually and carefully before you let them in, quickly locking the door behind them. You then have to unlock the door again to let them out. So, dear reader, on a busy summer night, when groups of kids numbering fifteen roam the streets and there are constantly six or seven cars on site and at least five people always entering and exiting the store, don't expect service with a fucking smile.

While safety was apparently always the number one priority, this was just *doublespeak*. My company drilled it into me that the door must <u>always</u> be locked at night. No exceptions. If someone came to check up on you and it wasn't, there would be hell to pay.

One bitterly cold night the door decided it wouldn't even close, let alone lock, so I called the night retail area manager. After I explained the situation to him, he asked me what the problem was. I explained again, adding that, besides it being colder in the store than the fridge itself, any psycho could just waltz inside whenever they wanted, which also meant that I couldn't do any work, or even go to the toilet, in case someone walked in and started racking stuff. Instead of sending a security guard and getting the door fixed, the manager told me to continue on as normal. He said the door wasn't 'high priority' enough, and would get fixed on Monday. Again, in the words of Orwell, 'Doublethink means the power of holding two contradictory beliefs in one's mind simultaneously, and accepting both of them.'

All the tyranny was making me a bit loopy at this point, so I decided to decompress, venturing out to catch up with some mates I hadn't seen for eons.

One of these mates goes by the name of El Diablo. Well, not literally. His actual name is José, but El Diablo is his nickname, for reasons that will become apparent in time. We've known each other since Grade 6. He's an excitable fellow. A verifiable lunatic, really, but with a fierce and unyielding heart. I love him as a brother.

We were at El Diablo's for one of his dad's famous barbecues. Many of my mates were there, as was most of Diablo's extended family. His dad, who had emigrated from Chile, presided over these affairs, cooking mounds of meat covered in homemade chilli sauce and gregariously laughing and singing with everyone, eventually playing old Beatles songs on his guitar until Chilean wine, and later, multiple pisco sours, would make even that an impossibility.

The slow-falling sun smeared firestains across the yard and all its smiling faces, while lighting the many bottles of Chilean red like long fat rubies as we sat around talking. Diablo had just returned from a three-month jaunt to Chile and he entertained me with stories with his usual aplomb, which was just the thing I needed. In return, I was telling him all about my new job and its endlessly Orwellian rules. Diablo's father overheard our conversation and chimed in, giving me a first-hand account of Pinochet and his fascist dictatorship.

'Fuck them all, the pigs!' he cried, gesturing wildly with his glass of Malbec.

The glorious smell of roasting meat, garlic, chilli and red wine was wafting through Diablo's backyard, and as I looked around at the laughing tipsy faces, I realised that this was what I'd been missing: simple, joyful, daytime human interaction, with no cages, threats or waking to a darkened world every evening. My own Malbec was going down easy, and for the first time since The Prom with Stevo, I felt human.

We sang songs, backed by a legion of guitars, into the early hours of the morning, until finally it was time to leave.

As I finished the last round of goodbyes, I headed inside to find Diablo.

'Dude, I'd better get going.'

He grabbed me in a fierce bear hug.

'Fuck, look after yourself, buddy! From what you've told me about those bastards you work for, it sounds like it'll only get worse. And I should know! We only met 'cos my dad got the hell out of Chile when he could, away from those Nazi fucks. Don't let them put you in chains too, brother.'

'Fat chance of that,' I muttered, skolling the last of my wine.

'You know what, we need to go out sometime soon and party, so you can forget about your shitty job for your fascist dictators!'

'It's a date.'

CHAPTER 11

THOUGHT POLICE

Around this time, a new diktat emerged from the wise cadres at the Politburo. As a card-carrying member of the nocturnal servo proletariat, I was required by my company to show up early for my weekend shifts and take the dip readings outside – at 11.45 pm at night.

This involved me donning a massive fluoro jacket that hung down to my ankles and kneeling over several small round holes in the middle of the forecourt to lift up long petrol-covered rods to gauge the levels of fuel in the large underground tanks. It's not that I necessarily minded doing the job (the night air – minus the petrol fumes – made a nice change to the stuffiness of inside), it was just that where the dip tanks were situated was directly in the path of entering traffic. And while I was lit up like a fluoro Christmas tree, the only protection between me and the screaming Commodores that would fishtail wildly into the servo was a bright yellow recycling bin. This was from the company that continually drilled into me that safety was *always* the first priority.

As this cheerful hypocrisy continued to make itself clearer and clearer, I decided to run an experiment. For one entire Saturday night, I made a commitment to test the bounds of fallacious *doublethink* and follow company policy to the absolute letter, regardless of the circumstances.

I showed up ten minutes early that night, a happy little apparatchik. I was cleanly shaven, my freshly ironed shirt tucked in, collar buttoned down and a pleasant odour and disposition emanating from my every pore. For the glory of Mother Servo, every customer was pestered with a myriad of company-suggested questions about chocolate bars, chewing gum and rewards cards. This was even if they were deep within MDMA's silken embrace or almost comatose as three fifths of scotch sloshed ruinously around their brain.

But before this could even happen, they actually had to get *into* the store. And tonight, it was a company-approved finger on the button. I scrutinised each and every customer who approached the locked door, with some of them banging their head into it, being used to it opening so effortlessly for them on so many nights in the past. I'd stand there and carefully study the image on the security monitor, even the regular customers, searching each individual for weapons, hidden features or anything else that looked slightly suspicious. This took a full minute before I could be sure of my safety and allow them entry into the store, ignoring their shouts and flailing arms as they jumped on the spot, trying to figure out what the hell was going on.

Unfortunately, according to company policy, anyone who arouses suspicion of acting violently, even in a verbal manner, shall not be allowed entry into the store. So, once again, I followed it to the letter. Angry throngs of customers bleated at me through the glass windows, sliding their fingers across their necks as they described in lurid detail the things they'd do to me when they finally got inside.

This was, according to company policy, an 'incident', and thus required the full five-page completion of a detailed incident report. Thus, for the next four hours, I filled out countless incident reports, detailing in colourful, precise language the threats hurled at me every few minutes.

Just as I ran out of forms, piling them all on the long-suffering Barb's desk, the milkman arrived, and I rushed into the fridge, not allowing him to leave until I had taken then confirmed each product's temperature and checked all the individual dates of hundreds of cartons of milk. He snorted at me in surprise, and said he had deliveries to do. I told him that, in that case, I was unfortunately unable to accept the delivery, for I could not, under company policy, guarantee quality assurance for my store and its valued customers.

After the swearing, red-faced milkman left with the same milk he brought in, I turned my attention to the store itself. As I checked the temperature of the ice cream freezer, I noticed it was 0.2 degrees Celsius colder than it should have been, thus not falling into the official temperature range allowed by official company policy. The only option, conceivably, was to empty the entire freezer, placing all the now-dangerous ice creams into the larger freezer out back.

After I completed this required task, I went to sit down, trying to take my mind off all the commotion with a Chocolate Big M and a large sausage roll with sauce. The only problem was, I wasn't allowed to eat them, because I couldn't pay for them. I couldn't pay for them because, under company policy, I was not allowed to serve myself, thus I would have to go hungry for the next six hours. You can kind of guess how the night went from there.

With work, uni and sleep duelling for my time, friends were left behind. I went through a three-month period where I saw my best mates maybe twice, max. And that was when they came in to get petrol. Even though I'd been working there over a year, they'd ring me during the day and ask me what I was doing Saturday night, week after maddening week.

There was this leather-jacketed hot chick I liked at uni. Bohemian-sexy-hot. Like kooky, short skirt, fish-net stockings, op-shop, groovy beads, writer-hot. Throughout the year we continually made eyes at each other in Advanced Non-fiction. But eventually, the eyes stopped, for mine had become puffy, pink and hidden behind the slits that only opened for loud noises, threats and Commodore high-beams.

So, action-less, sleep-deprived and rapidly sinking at uni, I toiled on. To make things even more interesting, my company upped its requests that I work at several neighbouring servos, due to their severe staff shortages. Every second night I was working at a weird and wonderful new store where the scheming locals would eye me with suspicion, before sneaking down the aisles and shoving as much chocolate as they could into their pockets before running out of the store. By this stage, the bitterness I'd accrued against my company and my lack of sleep made me silently cheer them on as I feigned a token shout of surprise for the cameras.

And as the night leaked into day, I drove home in a daze, ready to shower, dress, then head back the way I'd come for hour upon hour of uni. I'd sit in a darkened lecture theatre as the drone droned on while pointing at his projections. My head would droop precariously forward, just like Beavis's, only to jolt back up, scaring the crap out of the poor person sitting behind me. After it happened a few times, I learnt to sit in the back row.

I drove home from uni feeling like I was flying a spaceship. The lights would swim by as my full-blast air-con and the ineffective caffeine in my bloodstream struggled against the implacable, cloying nemesis of sleep, waking me up only with images of horror Transport Accident Commission advertisements from years gone by. I'd be thinking of assignments due, lecture attendance requirements, and the faint imprint of a life I used to have, but at the same time would be remembering the 3 am Gumbleton who spent a solid hour lecturing me on why Superman was more of a people's hero than Batman.

The utter antithesis of a people's hero was Nadine. She was, fittingly enough, the first ever manager of our servo, long before my time – kinda like how Stalin was appointed the Bolshevik representative to the Executive Committee of the Petrograd Soviet – growing to become the perfect suppressive instrument of the company. Nadine's ruthless vision of 100 per cent compliance rapidly elevated her through the ranks to become a retail area manager running fifteen or so servos, and mine, for about a two-year period, was one of them.

I lived under the reign of many area managers, and Nadine was the only one who frightened me. She had the power to make people disappear, just like Josef did during the great purges, and she did, with an almost casual consistency. She never liked my first store manager, Terri. Soon enough, Terri was shunted off to St Albans. To those who know Melbourne, St Albans is the Siberia of servo land, a forsaken, windswept hellhole of burned rubber, sirens and beady-eyed desperation.

I remember when I met Nadine for the first time. Thomas, the 'Disney On Ice/Tambourine on Your Front Lawn' guy who came to relieve me in the mornings, had just started his shift and was putting away the smokes. Hearing the store's phone ring he picked it up and was berated by Nadine to 'get back to work and stop mucking around with the cigarettes'. In shock, he looked up and saw her standing on the forecourt, on her mobile phone, spying on him, *Scream*-style. Turned out that she lived just around the corner and liked to . . . check up on things.

She walked into the store, looked me up and down with apparent disappointment, and launched in: 'So you're David. Tell me about this drive-off you had last night.' After I informed her, she accused me of sending the wrong information to her. I, being fairly sure I was right, and hungry, tired and in a bad mood by now, said that she was welcome to check her pager and we could see. She stared me down for a full five seconds, before dismissing it as unimportant and finding something else to berate me about.

From that day on I felt like I was a marked man. She didn't forget things. She would come in and spend hours watching videos of us all, dropping unsubtle reminders of conversations we'd had with customers and minute indiscretions we'd made. She was essentially the Thought Police.

After running out of things to criticise me about, she focused her darkness on Nellie, who turned pale as Nadine asked to see her in the back office. They were in there for fifteen minutes, speaking in hushed tones. Eventually the door opened and Nadine walked out with a smile on her face, neglecting to say goodbye to me as she exited the store. I went into the office to find Nellie, shaken.

Nadine had told her that because she had changed the videotapes of the cameras twenty minutes late, she was under suspicion for stealing stock. She said she would be looking into it and informing Nellie of her decision. But immediately after the accusations, she morphed into happy-Nadine, congratulating Nellie on her recent Long Service Award and asking how her kids were doing.

Nellie was bewildered by the flip. 'She just . . . changed, David. Just like that. Like it was the smallest thing in the world.'

Over the years, dozens of workers had made official complaints about Nadine, but not one of them ever stuck. Eventually, though, she pissed someone off who was higher up the corporate ladder than her and was gone. Like a puff of noxious smoke, clearing away to reveal a perfect blue day. The news spread like wildfire. Spontaneous parties were thrown in servos all over the western suburbs of Melbourne. It was like '53 in Soviet Russia.

But there were other ways to keep us in line. Sometimes, my company just decided not to pay me. At one point I was owed over $1200. It took me two months and five attempts ringing

Corporate before I was finally paid what I was owed. This would happen to almost everyone, at regular intervals. So whenever we got paid too much (which happened more than you'd think) we took it and shut the hell up.

Speaking of money, the one productive thing I had been doing for some time was saving it. With the perpetual triad of work, uni and precious sleep, I couldn't have spent it even if I'd wanted to, but I was definitely squirrelling it away for something, it was just that, at the time, I wasn't sure what. Several weeks later, though, I acquired an answer to a riddle that had been forming in my mind.

It was the kind of cool autumn afternoon that crept inside your clothes and flapped them skyward, blasting your exposed skin and making you shiver in expectation of another Melbourne winter.

I was having a morose lunch at uni with Paolo. We were sitting outside, the wind whipping our faces. I'd thought it would wake me up for my Fantasy Writing for Children class that was up next. As I sat there, sourly sipping my triple-shot latte, Paolo gave me a look of concern as he sucked up his noodles that were waving in the blustery wind.

'You look like a man that has lost his dream, my friend.'

'Has anyone ever told you you're an extremely perceptive individual, Paolo?'

'Many times.'

'Well, what do you think my problem is then?'

He stirred his noodles with his chopsticks, then knitted his dark brows together in a measuring gaze.

'I would guess drugs. Or maybe a woman. Many men have sacrificed their health for a woman. Fought wars, even.'

'I do have a mistress, Paolo, and she is unforgiving.'

'Many women are. But why do you stay? What does she give you in return?'

'Money.'

'This is an interesting relationship.'

Realising that he was taking me seriously I apologised to Paolo and explained, 'It's my job that makes me like this. I work in a petrol station at night and it's badly affecting my life in other areas.'

'But what is so bad about a petrol station? How does it ruin your life? It seems a bright and happy place.'

'Do you have a car, Paolo?'

'No.'

'How often do you go to petrol stations?'

'Not very often.'

'They are evil places, Paolo. Evil.'

'By the look of you, I believe you. But why do you stay?'

'For money. I'm trapped. Held in place by the horrible lights and the horrible people that sap me until my resolve is gone and I do as they wish.'

'This is truly a horrible fate, my friend.'

'It is.'

After lunch, I didn't go to Fantasy Writing for Children. Instead, I went to sit on a park bench and have a think about things. I was failing three subjects and I was fast-approaching dangerous territory in another two.

All through my school years, I'd succeeded with ease. Breezing through everything thrown at me, yawning through Year 12 exams before getting into my desired course with plenty to spare. And yet here I was less than three years later, almost kicked out of my course and looking forward to a lifetime of button-pressing and freak-watching. I imagined myself many years from now, wrinkled and bitter, a cantankerous clone sighing angrily at underage kids asking for smokes, before warning them not to fuck their lives up, lest they end up trapped in the very same halogen hellhole I found myself in, as tears brimmed behind my dull eyes.

As I sat motionless on the bench, I let my thoughts sift through my subconscious. Inside my head was an image of all my wispy childhood dreams – a wife, kids, dog, a happy family, nice car and a successful job as a journalist – all herded by a massive leaf blower into a gargantuan oven, where they roasted until they were only the sad grey smoke that floated out the top, drifting away to those living happier, fuller lives in the warm kiss of the sun.

I must have been there for a while, just morosely ruminating as dusk stole over the day, the wan sun painting everything with a shimmering haze.

An obese woman struggled along the path, her every step an increasingly torrid battle. In her left hand she clutched her twisted pack of Peter Jackson Gold 30s with a kind of mad paranoia, making sure everyone knew they were hers. She stopped to retch, heaving with something more than mere smoker's cough. As she began to breathe properly again, she raised her head and locked her fierce eyes on mine, jolting my gaze away, back to the comparative safety of my shoes.

In a fluttering of brilliant black, a gargantuan-sized crow settled on the rim of the metal bin, merely two metres from me. His giant claws clicked against the steel as he searched for discarded food, eyeing me with suspicion as he circled the hole, poking his beak in but emerging with only a plastic bag for his troubles. Hoping for better luck somewhere else, he flew off into the death throes of the now fiery, triumphant sun as the uni campus slowly emptied into loneliness.

I sat there watching the gilt-edged clouds as the yellow yolk desaturated into grey, and made a decision. Uni and work didn't mix. So, I decided to fly away myself. To defer. And go to Europe.

I reckon, deep down, that's every 20-something's solution to the problems life throws at you: leave the country, get drunk and go exploring. Plus, it seemed that that's what all the good writers did. Go to Paris; get secret writer injection; craft bestseller. It seemed perfect at the time. My plans were further set in stone a few weeks later when I received a lovely note saying I'd actually been excluded from my uni course for failing so many subjects.

My new dawn had arrived, Europe here I come. But first I had to keep saving, which meant I was still under the company's complete control.

CHAPTER 12

SATAN'S CHARIOT

Speaking of control, one method I've yet to discuss is arguably the most effective fear-based control mechanism of all. It stems from a well-known servo phenomenon.

Glance around the console of any service station and if you look hard enough, you'll see dozens of combinations of letters and numbers. Scrawled over every scribbleable surface, they are tattoos in time; the result of fresh-risen panic as the console operator failed to find paper quick enough to scribble the numberplate of a car that was rocketing toward freedom with a free week of Premium Unleaded.

Whenever you tell someone you work in a servo, the first thing they ask is whether you get any drive-offs. People have always been intrigued by the phenomenon of filling up and simply driving away. I've never done one and I never could, now that I'm fully aware of the absolute shit storm that blows in every time the fleeced company seeks its pound of console operator flesh in lieu of lost revenue.

Cretins who drive off just don't understand that it affects no one except the poor sod stuck behind the anti-jump wire. All I'll say about these criminals is that I hope our laws one day mirror those in the US, where in some states a drive-off is a felony offence and is punishable by an immediate loss of licence, and either a $10,000 fine or one year in jail. Problem solved. Tyre spikes are good too.

We rarely had drive-offs back when I started. Maybe one a month. Back then, people drove into the servo, got out of their cars, picked up the pump and it started pumping petrol. I could have been in the fridge, on the phone or reading *The Age*'s sport section while on the crapper, it wouldn't matter; you were actually allowed to put petrol in your tank, before I had even laid eyes on your vehicle. There *was*, of course, a shrill-sounding authorisation setting on the system that would require us to approve, with a press of the touchscreen, every single car before it filled up, but it could easily be switched off, and was by those who knew how.

If it wasn't switched off, you'd always know instantly, because as soon as you walked into the store to start your shift, a piercing, high-pitched beeping would shriek from the console, with some helpless newbie holding his tenderised head in between customers. A few quick screen touches would remedy the situation and make the poor trainee smile in the newfound silence like he'd suddenly discovered Zen.

But as often happens when you trust people, they start stealing stuff en masse. Customers figured that since we were practically offering, they'd take their petrol on us. I didn't mind, really. My company wasn't one of those dodgy places that made employees pay for drive-offs, so what did I care?

The Stasi, on the other hand, *did* start to care.

I came in one night to the familiar shrill *BEEP BEEP BEEP BEEP*, explaining to the new guy how to deactivate the system as I headed to the toilet. While emptying my tank, I kept hearing the infernal sound, confused as to how pressing two options on a touch screen could really be that hard. I came out with a sigh and went to shut it off, when the machine asked me for a 'manager's password'. Not good.

Soon enough, every shift was like when your six-year-old nephew gets a new toy that makes all those ray-gun noises that he loves so much he keeps pressing the trigger until you want to pulverise the damn thing. We gritted our teeth and soldiered on, but soon enough, even our regulars wanted to hack their ears off with a rusty breadknife and stuff them into their ear canals in the short but interminable ninety seconds it took them to get their sugar, caffeine and nicotine, and get the hell out.

Now, under the new regime, you had to look at the car and make sure it had a numberplate before you pressed the authorise button. When it was busy and four people picked up a pump at once, your head was assaulted by an orgy of beeps. Everyone was stressed. On edge. Ready to snap. Smiles turned to sneers as The Beep turned everything to shit.

You'd come into work to relieve someone at the end of their shift and you would sense a kind of electric current pulsing through the air. Looking at your fellow employee, you saw that the demons caused by The Beep had burrowed deep. Your co-worker's hair, normally kept in a neat comb-over, was wild and untamed, while their shirt, normally tucked in tightly, was billowing free as they paced up and down the console area like a caged leopard at the zoo, greeting each approaching customer with a forced smile, all teeth and tension, flinching as the sound they'd rapidly come to despise sliced through their synapses, jangling their brain to mush.

It was their eyes that were the main giveaway. They seemed to have an unnatural glint to them, that animal eye-shine suggesting madness was waiting on a hairline trigger, yearning for the chance to explode. Instead of a normal greeting, they grunted, sizing me up – their fight-or-flight response leaning precariously toward the former – before relaxing and tell me they were 'having a bit of a rough night'.

Again, it probably seems like I'm overstating matters, but do you know what it sounds like when your hangover is interrupted by a shrill iPhone alarm clock less than 30 centimetres away from your ear? Imagine it several octaves higher, infinitely louder, faster and a hell of a lot more insistent, going off every fifteen seconds for eight solid hours. That's what The Beep does.

We weren't happy with the situation but in the end, we realised that we had to make sure people were going to pay for their petrol. And gradually, people got smarter. It happens sometimes, I'm told. Evolution 'n' all. Plus, they got pissed off about rising petrol prices. They became devious, cold, calculating. Veritable MacGyvers. Realised that if you came in with different numberplates then how were we to know that it wasn't theirs? They used to love giving us the wave before they drove off. So, one fateful day the order came from above: numberplates. All of 'em. Written down before a car is allowed to fill up.

Now, I worked graveyards, which, compared to the daytime, was not too busy customer-wise. It got a little busy in the mornings, but I could handle the flow pretty well. I wasn't looking forward to the whole numberplate thing, but I figured I'd survive. Spare a thought for the poor sods who worked during the day. Especially on a Tuesday when petrol prices were at their lowest in the weekly price cycle and the forecourt was a flowing river of angry steel. Thirty cars, dozens of people in the store, the fridge empty, a tanker driver wanting a pen, a broken slurpee machine, and a nervous underage kid trying to buy smokes, all punctuated by The Beep. Numberplates were impossible, we all agreed. Stuff 'em.

My company didn't like this. Apparently, if someone decided to steal petrol, this was our fault, even though they had cameras recording everything that happened on the forecourt. An angry note in the diary was only the start of it. If you had a drive-off, you had to enter it into the system; print two receipts and file one with your shift paperwork; fill in the details on a long-winded official police offence form; fax it off to the cops; fill out the drive-off information on your shift report; put the details on the weekly store drive-off form; inform your store manager; then page the details of the drive-off to your retail area manager. And if you didn't get the numberplate – watch out.

Soon, staff started disappearing. Not the usual disappearances of people who couldn't hack the job after one shift; these were people who shouldn't disappear. Good workers. Shunted off to servos at the ends of the earth or oblivion itself. Colourful graphs were again posted up on the store noticeboard from head office to show us how much we had disappointed them. Drive-offs soon became a thing to fear. You could have had the best shift ever, full of smiling, considerate customers until right before you finished when some petty criminal sped off with your good mood. You then had to stay for an extra fifteen minutes filling out all the goddamn paperwork, and afterward watching your mobile and store diary nervously for signs of your coming punishment.

But not even this was enough. We'd eventually got into the groove of acting like machines – scanning all of the approaching numberplates and getting them down on paper with The Beep driving us toward thoughts of bloodlust – but more and more people had figured out the stolen plates thing. The Ministry of Love's angry graphs were piling higher and higher, and were now personalised for every single employee. I swear I can sometimes look at a numberplate in my suburb these days and tell you who is driving the car, what cigarettes they smoke, and how long they take to pump their petrol.

I discovered the company's next decree as I headed into work the following weekend. Ralph was standing behind the register, legs astride like an intrepid explorer as he studied the forecourt with a bright yellow pair of binoculars.

'What the hell are *they* for?' I asked, deep in my heart already knowing the answer.

'Every suspicious car must now have its numberplate checked against its registration sticker,' he said, looking rather at home with the binoculars in his hands.

'Superb. Gimme a go.'

'Okay, but the left one's a bit smudgy, so you might have to do it one-eyed if you want to see properly.'

Looking through my one eye I saw a close-up of a swinging soccer ball, hanging in a mini net from the rear-vision mirror.

Moving slowly to the right, I saw an extreme close-up of an extended middle finger, presented ostensibly for my benefit from the car's front passenger.

People didn't like the binoculars. And neither did we. It was a step too far. A brilliant idea and all, but you felt like the Terminator, scanning the forecourt for signs of potential danger. And you couldn't exactly waltz around the store with this massive pair of binoculars dangling from your neck, especially when you already had a duress pendant hanging off a wanky faux-gold chain. You looked like a mix between a confused wigga and a safari adventurer. Plus, considering my company went on about delighting its valuable customers, I figured treating them all like criminals probably wouldn't have the desired effect. I learnt to rely instead on the one thing that had served me so well up to that point – simple intuition.

You see, there are always signs that someone's getting ready to do a runner: downcast dudes in hoodies, drivers keeping their engine running, the passenger doing the pumping and, of course, always choosing Premium Unleaded. Some employees, namely Nellie, suspected every single customer, no matter how unlikely they seemed to drive off. I remember one night as a late-model Mercedes pulled up and Nellie checked the rego sticker with the binoculars.

'But it's a Merc!' I cried.

'How do you think he got it?' Nellie muttered, never taking her eyes off the car.

The company sent us many helpful notes on how to recognise and stop drive-offs before they happened. According to one memo, apparently 'ninety-five per cent of all drive-offs (were) completely avoidable'. While the memo had several helpful suggestions, like 'making eye contact with the individual' or 'noticing dirty numberplates', I had my own system.

To me, working out who's going to drive off was like playing that old game Guess Who? where, with the answer to each carefully worded question, you flip down your little squares of people until you're left with only one suspect. With drive-offs, the game would go something like:

'Do I have P Plates?'

'Yes.'

Flip down one square.

'Am I wearing a beanie?'

'Yes.'

Flip down three squares.

'Am I wearing fluoro hi-vis clothing?'

'Yes.'

Flip down five squares.

'Am I filling up a VN Commodore?'

'Yes.'

Game over! Play again?

To a service station attendant, the Holden VN Commodore is Satan's chariot. At one stage over 90 per cent of drive-offs from my servo were all VN Commodores. I think this is because it's the perfect bogan car. Once the flagship of Holden's range, it's a cheap but still powerful vehicle. For a few grand you could have something with a 5-litre V8 engine that produces 165 kilowatts of power. Great for doing burnouts. And screaming away from petrol bowsers . . . The VN is the calling card of the scheming no-hoper. The means of conveyance for all that is wrong with this world.

Your typical VNer is either full of nervous, aggressive energy, or is so stoned that you wonder how he made it to the servo at all. He will invariably be hawking narcotics or consuming them. Either bursting out of his tightly fitted shirt or so thin that his Dada or Fubu top is swimming on him. He will have a smoke tucked behind his ear and beady eyes that dart around incessantly, searching for police. Every time you see a VN drive up, a little warning light goes off in your amygdala, as either a screeching drive-off or four minutes of wailing about the recent 25-cent price increase on a packet of Winnie Blues is about to happen. The women are worse than the men. They have voices like a rusty chainsaw and a bogan glint in their eye that says, 'Don't fuck with me, sweet cheeks.'

It takes a while to get used to, the VN thing. When you start working at a servo, despite what your fellow workmates tell you, you find it hard to believe that everyone who drives a certain make or model of car can be inherently evil.

For me, the penny dropped one winter morning when a Commodore pulled up to pump six, one of our 'high-risk' pumps. Recognising that it was a VN and that it had tinted windows, the warnings I had received flooded into my head as I quickly scribbled down the numberplate. I watched apprehensively and then immediately relaxed as the driver emerged: a grey-haired woman with glasses and a knitted jumper.

As I busied myself serving customers and writing off stock, I looked up just in time to see her hang up her pump, give me the finger and hop into the car which then burst away, its tyres squealing. All VNers are evil. If not, the car will soon corrupt them. It has a sinister power that is not to be underestimated.

Although I forbade them to use my servo, many of my acquaintances were experts at drive-offs. One mate would give the console operator a sporting chance by giving him a hint as to his future actions. He covered up all the letters and numbers on his plate bar two: 'G' and 'O', which after filling up with Premium Unleaded, he then did, at breakneck speed, while giving the simpleton behind the counter a wave.

More and more often I'd get people come in to fill up on their way to work. Seeing that their car had no plates, I wouldn't authorise the pump, which would beep and grind its way into my brain as they marched in, demanding to know why their petrol wasn't pumping. When I told them they had no numberplates, they laughed and told me not to be stupid. It was only when I forced them to turn around and actually look at their car that they realised I was right. It usually turned out that their plates were stolen during the night by a group who would then use them to steal petrol from several servos. Kinda sucks.

What sucked more was when one genius once stole the plates from the car of one of the new guys we had working there and then filled up and drove off. The diligent employee jotted down the numberplate and faxed a drive-off form to the police, who contacted him soon after, asking him why he thought he didn't have to pay for his petrol.

But I take heart from the fact that not *all* the people stealing petrol were the sharpest tools in the shed. I've lost count of the number of times a guy (always a guy) would drive in, hop out, fill up his car with Premium Unleaded, grin, and then hop back in, waving as he drove off, ostensibly oblivious to the 'for sale' sign plastered on his back windscreen – accompanied by his mobile phone number. The cops even let me call him before they did. Oh, that was sweet.

The other type of drive-off is when people can't pay for their petrol. Whether they've forgotten their wallet, didn't get paid, or their significant other has spent all their dosh without telling them, heaps of people end up committing, in the words of my company, an 'insufficient funds drive-off'.

The thing is, though, most of them never come back and pay. Soon, my company started getting pissed off. Again. If they came into the store and couldn't pay, I was to show them a big yellow laminated notice full of legal mumbo jumbo that pretty much said: 'You are now in a serious amount of trouble. Give us money. Now. Or go to prison. And next time, don't mess with us.'

At the start of the blitz, people only had an hour to come back and pay before we involved the police. This was soon extended to a day, because the cops got sick of making threatening phone calls to bogans. I'm sure they do enough of that already. So, it was then left up to Barb to chase up the stragglers. She did this for about two weeks before she went insane. So, we had to act all tough when people came in and couldn't pay. Act like club bouncers with dead eyes and give them no options. Andrzej took to it like a duck to water. Soon some bright spark from head office suggested we could get the driver to phone a friend, who would give us their credit card details so we could enter them into the system for payment. People, quite rightly, told us to go and blow a goat when we suggested this.

Still, no matter what we tried, both sorts of drive-offs continued, and we were harried, threatened, warned. But how the hell can a console operator magically reduce the amount of people that leave their wallet at home, forget about their Netflix subscription, or simply decide not to pay? To this day it angers me that so many console operators spend countless hours worrying about whether they still have a job, simply because someone forgot something or they weren't able to read a tiny numberplate over 15 metres away, blocked by bowsers, trucks, an open hood, and cars that change and move forward every twenty seconds in a forecourt so busy it seethes. To this day, my company still hasn't found a way to get the money back from people who can't pay for their fuel. It kinda makes me smile.

If they were willing to accept something other than money, though, they might have had some success. People will try to offer you anything to pay for their fuel. Over the years I've been offered sunglasses, fake Rolexes, countless drugs, reverse hand jobs, a palm reading, and one woman who tried to pay for her Premium Unleaded with a large, lukewarm BBQ chicken pizza, with one slice already missing. 'I haven't poisoned it or anything,' she said, with a look of cold disdain that I would even think about turning down such a winning deal.

CHAPTER 13

SEX, DRUGS AND FLUORO PEOPLE

Okay, I should probably explain what fluoro people are. In Melbourne's working-class western suburbs, fluoro people are a regular phenomenon. Think shift workers who do long hours of manual labour. They include council workers, forklift drivers, picker-and-packers, construction workers, truck drivers and people who hack up giant slabs of meat in a cool-room.

The thing all fluoros have in common – besides not being burdened by any overarching sense of self-awareness – is that due to safety reasons they must wear high-visibility, fluorescent clothing at all times, whether a vest, polo shirt or a big bulky jacket. Fluoros keep their fluoro clothing on long after they finish work, so you can never tell if they're on a break, have just finished, or haven't worked in over a year. It's kind of like a glowing, radioactive second skin.

Comprising just about every nationality, but mostly male, fluoros are usually well-built guys. Strong as wild boars.

This fluoro guy came in one morning with his toddler son hoisted on his waist, and bought about $150 worth of groceries. After I squeezed them into a single bag at his request, he handed it to his son – who held it comfortably in one hand.

Their jobs are hard work, from what I can gather, because whether they're eighteen or sixty-five they always look like they've died then been resuscitated by forty-five seconds in the microwave. They run on stimulants and sugar. Most will stumble in at 2, 4 or 5 am and head straight for the fridge, grabbing as many bottles of Red Bull, V or Monster as they can carry, and then at the counter, several packets of smokes, strong ones in packs of 50s.

Two of them talking generally sounds like this:

'Ya want sumfin ta fucken eat bro?'

'Ah . . . fucken . . . Naah.'

'C'mon mate, I'll shout ya sumfin.'

'Nah, fuck it, it's all fucken crap here, we'll stop at Macca's or some shit.'

Perhaps not the most classically well-educated, they appreciated simpler things and, overall, they were some of my favourite customers: they were always good for conversation and they all had entertainingly bad jokes. My favourite: 'Do you want a bag?' (Them): 'Nah (grinning), got one waiting for me at home, mate.' They also wouldn't have known what a discount docket was if it was thrust under their nose, as it often was by their insistent girlfriends.

But their jobs just absolutely murdered them. You could see it in them all. You'd know when one of them had just finished a twelve-hour shift because they'd bound into the store like they'd been freed from a Siberian salt mine. It's hard to displace the surreal image of a massive man in a yellowy-green jacket festooned with coruscant silver panels skipping and whistling like Fred Astaire as he pirouettes past the Doritos, only to trudge in the following morning on his way back to work, pale and defeated once again.

There was a guy I knew who used to come in, this happy, friendly dude named Vinesh, who ended up resorting to a fluoro existence. He was originally a 9-to-5er and worked in IT. He'd show up every Monday morning for his petrol and newspaper, freshly shaven, showered and aftershaved, with not a hair on his well-combed head out of place. We'd share a joke and discuss the morning's headlines before he skipped off to his busy day. One Monday morning Vinesh didn't come in. When he finally did show up a few days later, he told me he'd been retrenched. Then he turned up early one morning in a fluoro-yellow polo shirt and I knew his days as a productive go-getter were gone.

The descent came quickly. While Vinesh had smoked before he joined the fluoro ranks, it was only now and again, and only Peter Jackson Ultra 20s (which, before the government took away the numbers telling you how much tar and nicotine was in each pack, were 4 mg). Slowly, as he toiled away at his new job as a forklift driver and became one of the damned, he moved his way up, buying cigarettes that were progressively stronger every month. Three months later he was smoking a 30-pack of Peter Jackson Reds (16 mg) a day, along with his two bottles of Red Bull. He was also about 10 kilos heavier and instead of laughing at my lame jokes – which he used to do with a friendly flourish – he would now blink at me absent-mindedly, the bags under his eyes shuffling unenthusiastically as if he was trying to remember the old Vinesh.

Gav, meanwhile, worked at a bakery near my servo. Not an official fluoro in terms of the uniform, but he qualified because of the shitty hours and manual labour. He rode in every night on his BMX at 1 am on the Red Bull Run, leaving with eight bottles for himself and the other bleary-eyed bakers. I always put them in two bags so they'd balance out on his handlebars. We got chatting one night about the relative crapness of our jobs and I slowly got to know him.

Gav's all-consuming hobby was stealing people's lighters. Every night he was out, he'd borrow someone's lighter and then casually walk off, adding it to the growing collection in his pockets before finding a new victim. His life's ambition was to get to 1000. He even showed me a photo of his colour-coded collection on his phone. It gave him so much joy that I gifted him the moulded plastic lighter display holders to store his rainbow booty in style.

One night Gav came in and said, 'I'm thinking of trying some speed. I spend so much money on this stuff and it wears off too fucken soon!' he laughed, gesturing toward the Red Bulls on the counter.

Two weeks later he flew into the forecourt, skidding sideways with his bike scraping along the asphalt, jumping off and through the opening glass doors all in one fluid movement.

'*Fuck!*'

'Good stuff?'

'I feel like Superman. Times TEN!'

Gav worked on speed for five weekends straight. Ironically, he ended up buying more Red Bull than ever before. Soon, though, he decided to work weeknights and use the speed, along with various other goodies, for going out instead. While on the plus side, his lighter-thieving reached heights heretofore unfathomed, I'm going to skip the decline and move straight to the nadir.

It was just after 2 am and Gav slid through the parting glass and raised his arms to the fluorescents, slowly turning to look at me with a grin so wide and slyly manic it was pure Joker. He danced around the store, gettin' crunk to the Lil Jon beats booming, screaming and crunching from the open windows of his idling car. A VN Commodore, naturally. Topless, his new pec tattoo of Snoopy was on proud display as he 'boom-chicka-boom'-ed to the slurpee machine and poured – his mouth replacing the slurpee cup – a liberal serving of both flavours. While he'd always had a thin pale frame, his shoulders now slanted noticeably downward from his neck, his arms were bony twigs and his collarbone was a deep white valley.

'Wassssssssuuuup Daaaaaaave?' he enquired, after swallowing the slurpee and shivering, then turning the shiver into a jerking dance that bounced along with the autotuned beat pouring from his car.

'Fuuuuuuck, I've got *brain* freeze! What did I come *in* for?' He ran out and hopped back into his car, turning the music up even louder before he ran back inside.

'Gettin' crunk is hot, man. Like, really hot! I'm burnin' up like . . . like . . . like the *SUN dawg*!' He bolted to the back of the store, yanked open the freezer doors and jumped inside, where he sat atop several bags of ice and, well, chilled. After a few minutes, a respectable-looking, middle-aged woman came into the store and, of course, headed straight for the freezer for a bag of ice, where Gav was sitting, rocking back and forth, running ice cubes slowly around his nipples.

He crashed his car a week later. Wrote it off, in fact. He came in to tell me as his pill was kicking in. His missus broke up with him at the same time. She didn't want him around her daughter. Which was ironic, because she went out and got fried as much as he did, leaving the kid with her grandma (who was trying to adopt her for her own safety), then coming down and worrying about losing her kid. She'd always come in and cry about it to me, while stern old men lined up for their Sunday morning newspapers.

As you may have gathered by now, the other distinctive habit of fluoro people is their Herculean intake of narcotics. While the legal ones we sell are paramount to their professional existence, the others are the occasional icing on the stale cakes of their overworked lives. The problem is, of course, that it's *never* occasional – it's all the fricken time. I suppose twelve hours is a long time to be lifting 30-kilo boxes, and spending eight hours holding a 'SLOW' or 'STOP' sign could get soul-destroying very quickly, but I've never seen another socio-economic group so jacked up all the time.

Speed is almost always their first choice. Fluoro nectar, it is. I always knew what was up when every couple of mornings they'd burst grinning through the doors, bounding straight to the big bottles of Mount Franklin which they'd half-empty before getting to the register.

There was one fluoro I don't think I ever saw straight. Skinny, bald and intense, he looked like a short, evil Peter Garrett, the ex–lead singer of Midnight Oil and ex–Labor pollie. The first time he came in, he was all monosyllables and arts-and-craft-store eyes. I thought he was trying to stare me down, but soon realised he was, well, if not out of his tree, then at least residing somewhere in the fragile outer branches. He looked through me for a while then finally muttered, 'It's okay mate, it's just drugs.'

He'd come in and chat now and again, but always with that monotone voice and deadened stare. Completely void of any emotion. He liked his pills as well as speed, munching them like Pac-Man every night he was out on the town. He had a family, apparently, which was supported by the fact he had a baby seat in the back of his Magna wagon.

He finished work at six every Monday morning and would always pop in on his way home. I'd know it was him when I heard the trademark Magna engine whine and then the screech of tyres as he came to a rapid sliding stop. He was always flying at this stage, his pupils two black dinner plates as he bought three packets of smokes and some chewie for his jaw, which was working overtime. I asked him once if he was going home to finally get some sleep and he blinked, jittered and said, 'Fuck no, I've gotta take the kids to school.'

Porn is also high on a fluoro's list. I had this guy that would get jacked up on coke then come in and buy great piles of it. You know the ones sealed up in shiny metallic bags so you can't see anything except their eyes and *HOT GIRLS INSIDE*? He'd power in, coked-up to the eyeballs, head straight for the porn section and grab two bags which had three magazines in each. He'd rip them open and spread them out on the counter, chucking a $50 note at me. Any customers who wanted to be served stood back and waited cautiously as he eyed the pouting orifices, his body heaving up and down with need.

One night, as he was leering at the fleshy curves, he turned to a meek old lady who had been waiting to buy a box of tissues.

'Just put ya stuff here luv, I won't bite . . . *Hard*! Hah hah hah!'

She walked up uneasily, placing her tissues on the small part of the bench not covered by flesh.

'Have a squiz at that, eh! Bit different to when you were younger, I bet . . . Check out the box on that one!' he cried, using his trembling finger to trace the anatomy of a redhead leering seductively back through her legs at us.

'Hey! *Tissues*! Where'd ya get those?'

She pointed to the far corner of the store before thrusting a $5 note at me and shuffling back quickly to her Renault.

I should probably admit here that fluoros aren't the only ones with mindless jobs who like to dabble in life's little doggy bag of goodies. As the crazy nights crept on by, I started to care less about the worries of the daywalkers; I was no longer of their world. I figured that if the night was so interesting at just my servo, then there must be a whole fluorescent world out there, where nocturnal clowns danced in a blissful stupor, unfettered by the turmoil and expectations of the day.

With no uni to worry about and a home life I desperately wanted to escape, I began to drink. Seriously drink, with what I considered to be an admirably ruinous level of commitment. Recoiling in horror from Stevo's sickly sweet bourbons, I spent weeks with Andrzej's recommended single malts but soon gave up, after waking up feeling like the entire British army had walked through my mouth the previous night. In their socks.

I just decided to take drugs instead.

It all started – predictably enough – with a customer. I was struggling one wintry Sunday night to put away a delivery of stock and check the use-by dates of pretty much every item in the store. Perhaps sensing my broadcasts of silent despair, a random puffy-jacketed drug dealer, after paying for his five porn magazines and four packets of smokes, looked me square in the eye and said, 'I think you need some help, bro.'

'You can say that again.'

'I think you need some help, bro!' Vacuous braying laughter.

'Or you could just tell me how you're going to help me.'

He smiled at me as he reached into his pocket.

'Alright smartarse. Every night I come in here and you're lookin' like death, strugglin' to put all these boxes away. How about gettin' a bitta goey inta ya?'

'Goey?'

He threw a half-opened packet of Extra on the counter. 'Speed, bro. I got a bit left over in there so it's on the house if ya wanna give it a try.' He winked. 'Those boxes won't know what hit 'em.'

'Umm . . . Okay, thanks.'

'No wokkas mate. And if ya want any more you know who to ask, got it?'

'Sure thing.'

'But if anyone asks where ya got it from, ya don't know shit, got it?'

I tried on my look of bold self-assurance.

'No worries.'

As he and his noisy jacket swished off into the night, I locked the door and took the chewie into the back room to inspect the tiny baggie more closely. I opened the zip-lock top and brought it up to my nose. It smelled acrid and, if anything, like cat's piss. I left the baggie as I attended to deliveries and customers, but soon enough I was standing back in front of it, inspecting the off-white crystals as I pondered the two and a half pallets of boxes waiting mockingly on the servo floor outside. So I tipped a tiny mound of speed out and opened up my wallet for a note. Not having any luck there, I went to the register and borrowed a fiver.

It stung. And it cleared out my nasal passage something shocking. I walked up and down the aisles for a minute, waiting for something to happen. Just as I thought I was beginning to feel something, the doorbell rang and I rushed back to the console to unlock the door for a crazy-looking elderly man with a handlebar moustache and wildly unkempt grey hair.

As he strode toward me, I felt a massive surge start to rise through my torso. 'How are you tonight, young man?' he said.

'Ummm, f-fine thanks. How are you? Are you good here, tonight?'

'Why yes! Except I'm on a mission for my wife. She's sent me down here, the terrible creature, for a certain bar of chocolate, but I can't for the life of me remember what it was called. I was hoping you might be able to help me.'

'Yeah, yeah sure. What colour is it?'

'Well, I'm not sure, but I know it was two words . . .'

After ten minutes of walking around the store going through different chocolate bars, I eventually sent him away with a king-sized Cherry Ripe, a Time Out and a Turkish Delight, just in case.

As soon as he was out the door, I raised my hand to my heart, which was a Kalashnikov firing on automatic as my head unscrewed and my palms began to sweat. I rushed to the mountain of boxes with a manic grin plastered across my face and set upon them like some starving jaguar on a herd of cardboard capybaras.

In the space of forty minutes, I'd drained a litre and a half of water, put away the entire delivery and was ripping through the aisles, scanning the use-by dates like a machine jacked up to warp speed. After I finished all that, I started writing. What, exactly, I wasn't sure, just giddy narcotised nonsense: customers who annoyed me, random thoughts, the way the lights assaulted my eyes. The words flowed from my mind's glittering mainframe in a silver-blue torrent, as I chuckled, mad-professor style, at my newly acquired hyper-genius.

Interruptions enraged me as, halfway through a sentence I was convinced was pure Kerouacian, the doorbell or the pump authorisation would squeal, shattering my crystallised realm of pure creation as I rushed out to get rid of them and get back to my frenzied scribblings.

I ended my shift that morning a convert, with a tapping foot and thousands of meaningless words. While it was fun, I endeavoured to find better uses for this exciting new rocket fuel.

CHAPTER 14

SUNSHINE ON PURE WHITE SNOW

So, with a backwards body clock and a uni-less week stretched out before me, I forgot the sun-drenched town squares of Europe in the hope of instead discovering this visceral underworld that was calling my name, growing louder with every sinking sun.

I didn't have to look far; I simply followed the tune of the many pied pipers who stumbled in and out of my store.

As if fated, the next week El Diablo burst in on Saturday night and was soon babbling away to me in his excited methamphetamine chatter. He repeated his idea of a night out on the town and I readily accepted so we agreed on the following Thursday.

But El Diablo didn't have speed, he had coke. Beyond scenes of Tony Montana, dipping his nose into a pyramid of fresh Colombian before introducing the world to 'his little friend', I knew next to nothing about marching powder.

When it hit me for the first time, we were in a Spanish night-club. I'd just emerged from the fancy mirrored toilets, nose-height shelves thoughtfully placed inside the cubicles. The coke had just begun its numbing descent down the back of my throat, my teeth hugging each other fiercely.

There was a couple, two Latino partners, dancing what I think was the Bolero. They were luminous, bronzed and masterful. As he dipped and twirled her, she spun up each time – her red taffeta skirt swishing in a perfect circle – with a soft feminine smile but dark eyes glittering. She was mesmeric. Sublime. In fact, they both were, him proudly leading with masculine honour and her with lithe, lustrous femininity, their steps speaking a language I couldn't hope to replicate, but it didn't matter at all, because their sorcery was upping its amplitude with every passing second. My world had become a slow-turning carousel so coated in alchemy I hadn't yet located which galaxy it was jetting through. I was horny. I was in control. I was twirling reality on my middle finger like a damn yo-yo. The more I watched them, the more I realised I could do *anything*.

She twirled round to face me and beamed, brown eyes opening wide, and as she did, every cell in my body shot up to a new glinting place of residence. I was beaming. Surging. A power rose from the balls of my feet and up into the god-like network of my brain. El Diablo clapped his hand on my shoulder and burst into a fit of laughter.

'She is good, my friend?'

'She is beautiful.'

'Like the whispering of warrior angels, rising up through your soul?'

'Yes, I think that would describe it perfectly,' I grinned in still-rising dopamine bliss.

For the next two hours I fell deep into charged, passionate discussions chock full of import and mutual joy with imposs-ibly beautiful strangers, punctuated by several tag-team toilet trips, the answers to life's tough questions dancing attentively at the forefront of my whirring cortex, three steps ahead of every possible deviation in conversation.

Time was an anthemic hymn of surging now. I was charming, witty, decisive and giddy with goodness. Another line seemed like a great idea – the *best* idea – so Diablo joined me in the toilets, tipping out more celestial rocket fuel. By the time we headed back out, the glow emanating from my body felt like a low-grade orgasm. The angels were warriors in golden plate-mail, singing, breakdancing, siring new galaxies, their glorious voices rising in unison.

The only real problem, of course, is the rationale cocaine engenders in your amygdala, that the absolute best time to snort a big fat line of it is . . . not all that long after just snorting a big fat line of it. That, and the economics. Diablo shuddered at the 1000 per cent mark-up from what he paid back in Chile. Plus, the down is unpleasant, which of course leads directly to you wanting more to bring you back up. But what a high.

Anyway, I suppose I should tell you a bit about El Diablo. As you have gathered by now, he liked to indulge somewhat. It was almost as if his DNA contained a missing sequence completed only by the chemicals methamphetamine, cocaine hydrochloride, lysergic acid diethylamide and methylenedioxymethamphetamine. Drugs excited him, but somewhat paradoxically, they also calmed him. But not his poor suffering mother.

While his father had long given up on him, El Diablo's mother was a good Catholic Chilean woman, who still held out hope that her son would rejoin life's respectable 9-to-5ers. Eventually, she too was broken. She rose from bed one morning, awoken by the sounds of the television, only to come into her lounge room to see her son, two-thirds of his way through two acid microdots and several points of speed, rocking back and forth on the couch, laughing manically at *SpongeBob* on the widescreen.

His mother grabbed him by the shoulders and shook him violently, trying to rid him of the demons she was sure had possessed him. 'EL DIABLO,' she shouted, waking the entire family. 'VÁYASE!'

After we all heard this story, his nickname was set for life.

Diablo soon introduced me to more friends who were intent on tinting their world with all the chemical colours of the rainbow. Chief among them was Marty the Party. Well, that's what we called him anyway. Several years older than me, he was a gregarious and impressively industrious drug dealer.

Marty was often excitable, with eyes that lit up like a fire sale whenever the mood took him, which was often. He also worked in childcare, which was where he met Diablo (who also worked in childcare before he moved into bricklaying), with Marty's steady supply of narcotics securing the bonds of their friendship even tighter.

Marty loved his job. The kids at after-school care could never keep up with him, no matter how hard they tried. It was the perfect job for him really, running around, playing with kids, for he was really just a big one himself. One of the kids had ADHD and brought his Ritalin with him, which Marty 'borrowed' now and again, hoping for some kind of rush. Predictably enough, according to Diablo, it calmed him right down. He just sat there and chilled all day. When the mums came to pick up their kids, they were all asking what was wrong with him.

I often went out with Marty because he always said yes. The man had a pathological *need* to party. Even if he was bankrupt, had chronic fatigue syndrome and was starting work at 5 am, he'd still go out every time you'd ring him up and ask. I'm sorry to say that, in my midweek boredom, I abused this man's disease for my own personal gain. I couldn't help it – the bright lights of Pleasure Island were dancing across the dark water and I was swimming for it with everything I had. Marty loved his pills more than anything, and because he possessed a never-ending supply of them, he soon introduced me to the wondrous spells they cast.

I'd finish work on Sunday mornings, clothes changed and ready to go, emerging renewed from Plato's halogen cave. Magpies warbled as I strode out across the forecourt and oil-stained carpark, a free man painted by the first spears of light hurled across the horizon. They bleached the dark bowl of sky and its roving clouds a crazy laughing pink as I hopped into Marty's idling XR8, a small line of coke placed thoughtfully on a CD case in the back seat. Then that beastly V8 roared as we tore off for the 'recovery' clubs in Prahran.

Snatches of light bounced into my eyes as the car sprang over bumpy roads, and I grinned like a fool as Marty gunned down the entry ramp onto the freeway as my coke kicked in and the psytrance pumped me into a new stratosphere.

Psychedelic trance is amazing. Marty introduced me to it soon after I met him. It's the throbbing dissonant opus, the bacchanalian roar of seventh dimensional robot sex that was surely gifted to us from a vastly more advanced galactic civilisation. It contorts itself into sounds you never thought possible until you actually hear them for yourself. At the same time, it somehow calms you, tuning all your thoughts into higher evolutionary wavelength as it crunches and pirouettes its wondrous way through the aural holes stretched in reality's crumbling facade.

We pulled into a servo with the bass rumbling over the forecourt as the 9-to-5ers scowled at us in disgust. We smiled back, in massive aviator sunnies, and walked in and bought every packet of chewie Extra made and engaged the young worker in some chemical chatter, before strutting back to the car and screaming away from the pumps and into the full sunrise.

Sometimes we'd get to the club and I'd realise I'd forgotten my change of clothes. I'd simply head on in wearing my work uniform instead. The bouncer would see me standing in line, grin at my name badge with his two huge thumbs up, saying, 'Pump number three, thanks mate!'

Coke is nice, but there are no words for when MDMA fully subsumes you. I still remember my first time, in that dingy 'recovery' nightclub during the day on Commercial Road in Prahran. Marty and Diablo had been there so often they were part of the furniture, and seemed to know half the club, including the bouncers. I walked up the steps after them, giddy and nervous, the crunching beat, euphoric rising strings and satin vocals drawing me up like a rare earth magnet.

Marty surreptitiously palmed me a tiny white pill stamped with a dove and I inspected it in a dark corner before necking it and waiting.

Thirty minutes later, I'm standing around shuffling to the beat when I feel a spreading warmth and a rising clear haze. Colours jump out in visceral swathes. Beats and rising vocals slice through the din, clarion calls I can suddenly feel from my toes to my candy-coated skull. With each note, my thighs and forearms begin to buzz, my every pore opening like a satellite dish.

I'm softer, pliable and *happier.* As I think about it, I can't get that smile off my face. An internal sun is rising somewhere. The beat twists into a well-known anthem – Delerium's 'Silence' – transcendent and sublime. Her voice is rising, seraphic, and all at once something numinous slips into my synapses like third-date-sex with the girl of my dreams. I turn wide-eyed to Diablo – grin touching my ears – and lift my arms to the sky then yell pure, crystallised bliss at the DJ, who smiles back knowingly as he shuffles behind his decks.

Marty hangs a loving arm over my surging shoulder, takes me to the bar, and buys two VBs.

'VB? Really? You could have picked a better beer,' I grin.

'Trust me brother, this will taste better than the best craft beer you've had in your life.'

I take a sip and he's right; he's more than right: I'm drinking liquid nectar and every sip lights me up more and more. I feel an overarching sense of peace, of love, and I want to spread it.

I buy another drink and hand it back to the bartender saying it's for him. He grins and shakes his head as Marty face-palms and takes me back to the dancefloor.

My pill soon begins to earn its $35, mixing me into the insistent euphoria of the crowd that becomes one single entity as I pound and sway through the tumult. Huge muscle-clad Asian cokeheads hug each other, sipping mini bottles of water with petite grace. One of them hugs me, and tells me that he loves me, that he loves the world, and the music and well . . . everything. I sit on a red felt couch, giddy, thankful and diaphanous, splashed in brilliant lasers of crimson, sapphire, violet and gold, watching the DJ flip records in the air like pancakes.

I'm still coming up. I'm on a cosmic chairlift and I'm not even halfway up the mountain. The sun glints perfect off the snow. It can't get better than this, can it?

It does. I weave back onto the dancefloor, surrendering to the trance and my body dances *me*. Time loses meaning. An hour turns out to be seven minutes, and three songs an hour and a half. All my problems, worries, anger and doubts now seem hilariously inconsequential as I'm lifted to a place I never thought possible, surging above the mountain and into the skies.

Brilliant diamonds pirouette through the air; lucent buckshot from the burling crystal suns of mirror balls, painting the canvas of twisting smoke and threading through each other like ethereal strands of liquid DNA. I've swallowed heaven, and I turn to look at my neighbours, stomping in enraptured awe, as the cutest of them beams at me, before bear-hugging me and planting a soft kiss on my lips. I follow her like a puppy to the couches where she laughs at my work uniform.

She traces her fingers along my arms and each look into her face is a magic so strong I only allow myself small snatches at first. Her eyes are gold, wet with light. Her lips are perfect.

Her name is Claudia.

'Oh my god, this is your *first time*?'

I nod. I'm in love.

'Are you going to tell me your name, servo guy, or just grin in perfect bliss?'

I grin. In perfect bliss.

'Sorry, I'm Dave.'

'Well,' she says, fingering my name badge, 'I suppose I already knew that. Well, Dave, let's make sure your first time is one to remember.'

We dance, hours into minutes, then we're back on the couch again, glass of Amaretto now in my hand, Claudia on my lap.

'I want the sun on my skin!' she shouts, and we walk out into a fierce bath of light.

The club's roar fades and cars drone along the busy street. We walk hand-in-hand to a little servo to buy more chewie and water and everything inside is riotous plenty. A technicolour Narnia whose aisles I stride slowly in awe, enraptured by its shining jewels. I giggle, because I realise I've become a Gumbleton, but I couldn't care less. The console operator stares at me worryingly and I know his stare because I've stared it myself at so many flying customers at 3 am. But I'm on the other side now, and it's beautiful here.

She takes my hand and we float around the corner to the children's playground and her hair is gold in this apollonian sun. She skips ahead then stops, looking demurely over her shoulder to beckon me on. We reach the grass and she snow-angels on it, the sun caressing her midriff.

'This is orgasmic, I can feel it charging me,' she says.

'You have the most perfect bellybutton.'

She sits up, surveys me through narrowed eyes, then just as quickly switches to coquettishness.

'Is that *all* you like about me, Dave, my bellybutton?'

I kiss her in the sun and she traces her fingers down my neck, down my back, and with my arms around her waist, I feel like the half of a perfect whole. I know this is a chemical love but I don't care.

'So what was your first pill, Servo Dave?'

'A white dove.'

She laughs. 'What an appropriate beginning! The white dove of peace. Seriously, though, they're good pills. Very clean. They got a ninety-two per cent rating on Pill Reports.'

'You really know your stuff.'

'Everyone should know the stuff that goes into their body. Hmmm, it's gonna be hard to top a white dove . . .' She grins wickedly, 'but I'll see what I can rustle up.'

She takes out a baggie from her jeans and dips a long purple fingernail in, measuring a tiny bump then surveying it with slitted eyes before offering it to me, saying 'Sniff!', like Mary Poppins, then has one herself.

The pure MDMA powder soon kicks in as we watch two kids on the swings surf the sky in looping arcs. We lie and kiss in our magic bubble of bliss, everything else receding into distant sepia.

She smiles at me. 'Come.'

We walk away from the main road, down another street then up a leafy side street.

She gets her keys out and a short beep unlocks a midnight blue Lancer.

'I think I've had enough sun for now, Servo Dave. And so have you. We couldn't possibly get you burnt on your first magical adventure now, could we?'

A psyambient CD slides in and out pours impossible alien beauty. I sit back in the bucket seat and the slow strings, celestial vocals and deep African drums carry me back to Narnia.

A soft wet kiss lands on my cheek and she pushes me into the back seat. I kiss her, slowly, and stroke her soft springy breasts, and her black eyes sparkle hungrily as she straddles me. She whispers in my ear as I kiss her neck and she arches her back and shivers as I enter her and we rise and fall as the psy slow-climbs up the next mountain, and it's higher than the first, and the beat gets louder, deeper, and it CAN'T be this good. She's warm and dusky, and her hair drapes over me, and she moans and so do I because everything is perfect insanity, stretching out into endless clouds of lucent gold.

Sated, we lie in swords of milky sun, surfing atop the music and sinking back slowly into our separate bodies, and I watch the melody spread itself through the air to settle softly on the back seat.

'I can see the music.'

'What does it look like?'

'Like sunshine on pure white snow.'

Later. The sun deepens to tangerine and begins to sink over the tops of apartment buildings. We return to the club, our shadows long on the footpath in front of us. We slip back in and its neon roar is a crunching foundry of gods, hundreds of arms slithering Medusan under duelling swords of purple and green. We dance for a while but we're no longer feeling it. Everyone is sprawled dead-eyed on sticky couches, Marty and Diablo among them. I head to the toilets, where the floor is now an immoral landscape of urine, shredded toilet paper, spew, smashed glasses, and dozens of small, empty, clear zip-lock bags.

As the music thuds I catch a reflection of my haunted, elongated face in the mirror. Others enter and stare at me, and then at each other, and we all know the magic chairlift is heading back down the mountain.

I find Claudia and we crash on a couch, feeling hot, cold, heavy and cracked out, watching the new arrivals power up and on, arms raised in unison as they stomp through the folding labyrinth of laser. I squeeze an arm around her cold waist, trying to rub some warmth back into it, but my hands are cold too, and even that small movement is an ordeal.

I tell her I'm done. If work wasn't leering over the horizon, I'd find somewhere quiet where we could talk, wrap ourselves in blankets, smoke joints, and come down in peace, but my soul is heavy and I stand, feeling like a sack of burnt potatoes. She kisses me, then smacks me on the arse and gives me a crooked smile.

'You'll always remember your first time, Servo Dave.'

I find Marty and Diablo, standing by the bar, looking like two addled vagrants warming their hands by a bin fire. They stare back at me with huge funnel-eyes, but complete understanding.

After seven swirling hours, we leave what has become Madame Tussauds. Stepping out into a hazy evening, I feel more fragile and depleted than I have in a very long time. I am serotonin-less. It's thirty-one hours since I've slept, eleven since I've eaten. And it is four hours, seventeen minutes and twenty-nine seconds until I have to get ready for work. The bright circus box of colours floating on a sea of shadow waits for me; a sordid stage for psychos, clowns and Coke bottles. I trudge to the car like a man on death row, my legs sorry anvils.

We scrape home like wraiths, ancient maps of hours ago; cadavers in this carapace of leaking dream. Diablo weaves his station wagon between the steel-glass canyons that seem to whisper spite across our shoulders as we flee. We crawl over the Westgate Bridge and in the dark the city glisters like a scheming Oz.

As the streetlights and traffic signs shunt past, I'm a golem, wrapped in ghosts. Perforated. Sclerotic. Reading my obituaries with soaped-window eyes. They drop me home and Diablo turns around from the front seat and looks me up and down before remarking in monotone, 'Dave ... I'm more likely to crap out a penguin than you are to survive eight hours of work.'

I walk, head-down, straight to my room where I stare at the shifting alien wall and suddenly it's 11.17 pm, and I have to get ready for work.

I got dressed mechanically, slowly, and my work shirt scraped against my frayed nerve endings. My car crawled through the wet darkened streets, the sickly wash of headlights hitting the back of my skull as I rolled unwillingly toward that sad box of colours. It loomed wretchedly against the washed-out night, and as I stood by my car, I was Pinocchio, returned from Pleasure Island with donkey ears. I could hear the suffering of the world on the wind.

I trudged in through the auto-doors and the colours leapt maliciously into my eyeballs, which I kept downcast and away from the worker finishing their shift. After they eventually left, I looked out at the store and sighed. What goes up does indeed come down, and coming down in a brightly lit cage, looking out at muttering psychos and angry shift workers just didn't sit right.

My serotonin depletion was so fierce that sausage rolls tasted like cardboard and my face felt shame at my fake, fake smiles. I paced outside sometimes, not particularly sure what I was doing. The lights that shone from the canopy melted into the skeins of writhing shadow. Sirens pierced the dark and I jumped, thinking I'd somehow been discovered, running back to the flat discomfort of inside.

When I finally finished that horrible shift, I walked from the store to my car, stood there and looked up at the sky. I stared into this stunning sunrise, as pink and gold fire bloomed underneath a charcoal grey, hefting warm swords of light and gilding the furrowed curtains of brooding cloud. And standing there staring at it, a little panic rose up through my chest as I realised it meant nothing to me.

I didn't stop, though. Magic like that . . . once it wraps its silver wings around you, everything else is a dull facsimile of reality. Plus, the downs were actually pretty rare and only lasted a day or two, as my serotonin supplies came roaring back. Even then, the vastly changeable levels of neurotransmitters shifting through my brain meant I'd do some odd things while sideways. Reality felt like a hazy RPG game with no real consequences, so I wore name badges that didn't belong to me. Or I'd courteously wear a 'TRAINEE' badge, explaining in advance all the mistakes I was yet to make that shift to the customers. During a small break in between serving one night, I went to straighten the curled-up $5 notes in my till and the unmistakable whiff of cocaine wafted up to my nostrils. As I grabbed the wad of purple notes and held them close, savouring the sweet bouquet, I looked up at an elderly woman, who was staring at me worriedly.

Now and again, I'd line up all the cigarette packets behind the counter so that the government health warning on every single pack was exactly the same picture. Most customers didn't notice, but every so often someone would approach the counter and stumble as they saw over 100 cancerous mouths leering back at them. I went through a period where I would continually leave nonsensical haikus scattered throughout the store diary, weirding out everyone:

> Colours laugh at me
> Cars' headlights are very bright
> Customers can wait.

And if you've ever shown up to work a little fragile from previous festivities undertaken, you may have noticed all the evil things that only seem to happen when you don't possess the mental faculties to deal with them.

One Friday night around 1 am, struggling after some dirty hallucinogenic pills and no sleep for about forty hours, I felt I'd fallen through somehow to the Other Side of the glass, *Dark Crystal*–style, where only fiendish magic reigned. The Coke bottles and chocolate bars danced on the shelves like scheming broomsticks in *Fantasia*. Photoshopped starlets followed me with their eyes from magazine covers as I lurched around the store, Beyoncé pouting while Britney glared, incensed.

The shapes were swirling into the faces of brightly coloured beasts that began to march forward. Customers were too uncomfortable to stay in the store very long, so mostly I had peace. That is, until the delivery driver appeared, dragging a pallet of boxes on a metal trolley, catching the bottom corner of the glass auto-door and shattering it with a massive *pish*. He placed his hands on his hips, laughed and said, 'Damn, that's the second time this week.'

CHAPTER 15

THE HIGH PRIEST OF SWANSTON STREET

I spent the next week with a makeshift wooden door, and had to explain to every single customer what had happened. Because I had to do this close to a thousand times, I started making up outlandish stories to amuse myself. I told one naive customer it was a smash 'n' grab, while another wide-eyed fool believed my tale about a massive brawl between rival drug gangs.

As you might have expected with mixing drugs and work, there was sometimes a close call or two. I would often forget to lock the door while I hoovered a line of coke in the back room then strode out like Superman, only for my heart to do a triple somersault with inverted twist and 9.3 degrees of difficulty as I locked eyes with a police officer who hadn't seen anyone in the store and had popped in to 'check if everything was okay'.

Of course, things eventually got too much. I'd be finishing my shift Monday morning and my workmates would glance worryingly at my deathly appearance as I clattered the wrong coins into the wrong squares of my till and held up my shift report print-out like some long-lost Aramaic tome.

Your judgement goes, too.

Five hours before work, I was at a rave club in the city with Marty, assessing my options. I held out my two hands and realised I was Morpheus. In my left hand, the blue pill: 100 mg of 5-HTP to replenish my depleted serotonin stores from the previous eight hours of revelry and help me get through that night's shift. In my right: a 150-microgram tab of what I'd been told was high-quality LSD. Fittingly, this speck of noetic flame was coloured somewhat crimson.

I took the red pill and danced for a while, then on a whim I left the club, wandering Melbourne's streets as the electric fire started to foam up through my cells. I ambled down Bourke Street in the last light of day and in the gathering gloom, I sat on a bench, all filters of perception melting, my every molecule a nuclear reactor fizzing to life in unison with the city lights' amoebic bloom.

Sirens exploded from nowhere, and my eyes chased the battalions splashing red and blue and I stood painted in their twisting rays, thumbing off the night's safety. Shadows swam beneath the shouts, as Bourke Street's white fire sprayed Rorschachs of light onto KFC, but my mind shone fierce, the rarest of technicolour jewels, as my head spun untethered, laughing up into the sky.

I walked, unable to stop, now I was plugged in. On Swanston Street I came across a ghetto preacher, spewing out surprisingly eloquent brimstone at the cold Melbourne night.

With straggly swords of hair pointing down in diagonals and parallel universes sloshing in his irises, I watched him for twenty-five minutes and his material was always fresh, rhapsodic. He was plugged in, like me. He was the High Priest of Swanston Street, hurling truth from his milk-crate pulpit:

> *Wake children, WAKE! His amber eyes, they*
> *seek you with the swivel of a snake!*
> *You're drawn to his hiss, and his caverns of fire, like a*
> *neck to noose's GRINNING EYE!*

He was like Randall, but vastly improved; jabbed with plutonium and jacked up to the power of nineteen. The energy orbiting him was swirling and fierce. He'd gathered a small crowd, half of them with black dinner-plate eyes, and I felt them all, circling this cosmic plughole. I walked. And walked. And walked.

The city streets were virtual rivers of blood and gold. Time was a construct; waterfalls of green code in ones and zeros, but the specific numbers on my phone told me it was time to head back.

A taxi home. I gazed out the window at diodes of digital fire, as the car sluiced through the city's glittering circuit board, with me futilely trying to convince the driver from Addis Ababa how it was a metaphor for the overarching truth connecting all reality.

Home. A sixth-dimensional shower. Another taxi to work, eyes down, don't talk to Ralph, don't talk to Ralph. Focus every surging neuron on psychically pushing him out the door and into his Corolla.

People in block-shaped cars with block-shaped heads came in and bought boxes of Arnott's Shapes and boxes of cigarettes.

Square fingers extracted square wallets from their trousers to give me square money and I cried inside at the splendour of the symmetry of the geometrics, initially refusing their money because the poetry of the universe's mathematics was too sacred for my square till.

I told them their fiat currency was simply a sheet of plastic, a metaphor for stored energy, and that the world's true currency was love, rippling through the unseen, as much as it inhabited our own fierce hearts. They looked down their noses, studying me with square eyes; box robots scanning for signs of danger. I looked away, feeling naked under the beam of their stares. They left and I turned my gaze to the endless chocolate bars on the shelves, instantly tasting each of them in my mouth as I scanned my eyes over them.

Imagine trying to serve customers while feeling the meaning of life pulsing away like the birth of an alien sun in the shining caverns of your skull. Imagine it enveloping your brain with awe and childlike wonder as the dawn's golden rays bathe you in swimming arabesques of sacred light, making you shudder in hidden ecstasies. Then imagine the angry mono-browed man who bears a disconcerting resemblance to Spock from *Star Trek*, awaiting the return of his Visa card, a growing line behind him.

It was at this point that I convinced myself – for the final time – that drugs and work did not mix.

In my defence, at the time I really did believe I was an amalgam of Jack Kerouac and Hunter S. Thompson. Everyone needs a hero and back then they were mine. Plus every good writer needed a bit of life experience. Kerouac hopped freight trains across America, Orwell braved pneumonia in Paris and London, and Thompson pushed commercial-scale drug taking to the absolute limit in his search of the death of the American dream.

I didn't find anything as profound as that. I just got a bit fried and wrote about crazy customers on register receipt rolls in between stocking the fridge.

Turns out I wasn't the only one heading down the path of lascivious immorality. A year or so after I'd first met him, I noticed Ralph was starting to change. Although he was in his mid-twenties, he was always expected home as soon as he finished his shift. Just like Principal Skinner, his mother seemed to have this vice-like grip over him, always ringing to remind him to get milk, or to make sure he ate his dinner which she would dutifully drop off to him on his afternoon shift.

But soon I was regaled with tale after decadent tale: his first adventures with alcohol, his first foray into clubbing (he showed up at the local bogan nightclub in full suit and tie). Ralph was spreading his hedonistic wings. He became more relaxed at work, too. His shirt, previously always tightly tucked in, now billowed free as he adopted a more *laissez-faire* attitude to his appearance, the store, and the universe, beginning to bend some of the rules, just a little bit . . . Instead of stern disapproval, customers were peppered with wry grins and 'don't worry about it, mate's. He even began rebelling against several facets of company policy, as I watched on like a proud cousin.

Ralph became general manager of a suburban cricket side, which, with its alcohol-soaked nights of strippers and debauchery, led him slowly down the path toward blokedom. He'd show up at work unshaven and hazy from the night before. He started letting people off 10 cents here and there. A big step. The end of year cricket trip to Wangaratta was the capstone. He got kicked out of two different pubs in one night, so titanically shitfaced he was mumbling hieroglyphics.

And as I said earlier, I wasn't the only one on drugs. While I felt like joining in now and again, my customers always had me covered in that regard. Servos attract drugs, dealers and the varied lost souls like flames do moths. This makes sense when you think about it, because service stations are the biggest drug dealer of all. They simply pander to all the addictions our nanny state government allows us. Servos are cartels of sugar and nicotine, with petroleum and caffeine thrown in as afterthoughts. It's perhaps fitting that they stock so many products with the names *hit*, *fix*, *speed* and *rush*. You should see the smokers come in every morning, ripping the plastic off the packs and lighting up even before they get back to their cars. Of course, they'll end up leaving with two packs, thanks to good ol' buy-more-n-save servo deals.

Because of our public phone, we were the dealers' home base. I had two memorable regulars. The first I dubbed Coke Wanker, a smug and arrogant twat with bleached spiky hair who'd strut in sporting three chunky gold chains that hung down over his yellow Fubu jumper. He'd pull up in his Commodore with self-congratulating plates and buy $300 worth of crap, slowly unfolding a massive wad of bills to give me several crisp green hundreds.

He came in one night rather upbeat, telling me he'd not long got back from several hours at the cop shop. He'd just been done for trafficking. Said he was looking at nine and a half years.

He appeared quite nonplussed about the whole thing, but by then he was Tony Montana levels of lit, sitting on the ice cream freezer with his nose deep in a sandwich-sized bag of blow.

My favourite dealer was the opposite of Coke Wanker, more of a gregarious, bipolar Tony Mokbel. I saw him in a different car almost every week. He'd pull up in a brand-new Chrysler, a Jeep, a Monaro, an AMG Merc, even a customised SS Ute, on top of the world. Not on anything, just resplendent in his fiery robes of mania.

He paid for the fuel of dozens of complete strangers, gave random people lifts and even bought Randall both endless packets of smokes and a whole three-course meal of fast food, not leaving until Randall had eaten it all. 'You need some meat on your bones,' he'd chide, like a nonna in the kitchen. He did the same for me, bringing me Macca's and always leaving me with whatever change was left over from the fifties he chucked around like confetti, making sure it went in my pocket and not the till. Other nights he'd stalk in, dark and edgy, yelling to no one in particular about 'the fucken pigs and their dirty tricks!', steaming down the aisles like an Italian bull wearing red-tinted glasses.

Then there were the old drunken Aussie yobbos. One guy stuck out most. Shaun-Michael-English was his name, like some serial killer you hear of on the news reports. He was the quintessential old ocker alcoholic. As sharp as a bowling ball but always endearing. The first time I met him he was so trolleyed he couldn't even slur his speech. He swayed into the store on a sea of attempted greetings.

'G'day bl . . . oke . . . What *are* ya? I mean . . . uh, how (hiccup) ya . . . uh goin, then, bloke?'

'Okay mate, how about you?'

'I'm so hungry I could . . . (hiccup) . . . I could eat the arse out of a (hiccup) low flyin' duck.'

'A few beverages will do that.'

'Who (hiccup), me? Nah, I haven't had a cunt all night, drinkstable.'

He stayed in the store for an hour, munching roast chicken rolls, pies and random junk, sometimes making conversation, but mostly trying to remember his name. He'd totter every so often as gravity snatched at him from unexpected angles.

He waddled up and down the front of the console, picking up chocolate bars and trying to pronounce their seemingly exotic names. 'What's this one then, all (hiccup) f . . . ancy 'n' shit . . . *K I N D E R BE-YOO-NO*. Is this French or sumfin', bloke? If I eat it, will I turn into a (hiccup) frog?' (Cue hysteric drunken laughter as he rips it open and hands me the wrapper to scan.) '*Kinder Byoono*, ey? Well, let's give it a burl, bloke!'

He dropped a two-dollar coin on the counter to pay but it rolled back toward the floor, before he stopped it with his gut, bellowing at his ingenuity.

(Minutes later.)

'Stuff this, I'm goin.'

'Where you off to, mate?'

'Ta get me guns.'

'Guns? Why do you need guns?'

'To shoot that (hiccup) bitch of an ex-wife – and her snooty lover boy who flaps around like a dick in a fucken shirtsleeve. I told 'em I'd shoot 'em last night on the phone and they (hiccup) didn't believe me. Smart little prick told me to go ahead. So maybe I will.'

'Well, I dunno if that'd be such a good idea.'

'Why's that?'

'Well, you'd probably end up in prison.'

'Prison's fine ... Prison I can do standin' on me head. As long as they let me bring me dogs ... And me guns.'

'Well, I don't think they'd let you do that.'

'Really? (hiccup) Figures ... How about just me guns, then? I'll get someone ta feed me dogs while I'm inside.'

Other drunks would sit in their cars at the pumps, cans of Woodstock and Cola rested atop their steering wheels as they stared at me – for three hours. They didn't get out, or even move. Didn't do a single thing. Just sat in their cars. And stared.

There were mangy singleted lunatics on speed, so much that the foam left on the corners of their mouths had dried solid and they started each new sentence before they'd finished the previous one. They'd bluster about and talk this gibberish at me – for hours at a time – as the oily film on their shiny faces began to sprout the first rash of new amphetamine pimples.

There were the ravers – and, increasingly, just average Joes – who necked pills like they were Tic Tacs. They popped in at all hours, on their way to PHD or Hard Kandy; whatever was on that night. They swam around the front of the console, lost in a sea of dreams and beautiful things I now knew all too well. Their black eyes were like sponges, soaking in everything they saw with an enraptured awe and filtering it into a shiny fiction, as fifteen-year-old kids danced down the aisles to an imagined beat.

One night, a woman had been swaying around the store for several minutes before she approached the counter and beamed at me, spreading out four packets of chewie on top of our promo mat like a deck of cards. She was peaking so hard, her jaw was like a vibrator.

'Hi beautiful. How you doin' on this gloriously beautiful night?' she said.

'Dare I say, not as good as you.'

'Ha-ha, pay that one.'

'And dare I say, that chewie's gonna get a good workout soon.'

'And that one! I love to par-tay, but I should probably calm it down a bit,' she smirked, as she ripped open the spearmint and popped two into her mouth. 'I'm a dental nurse, see, and my boss, he's a, he's a . . . well, *he's a fuckin dentist ain't he*!? Har-dee-har-har!'

'I chipped a tooth and now I'm too scared to go into work today, because the bastard always checks out my teeth as some sort of visual reflex. I can't let him see me like this. I'm gonna have to go to a dentist . . . *Another dentist*! Hah!'

And then there was Hugo. Hugo was a plasterer. His hobbies, besides long stretches of inebriation, were collecting all of the graphic government warnings in sets of Winfield, Peter Jackson, Dunhill, and Benson and Hedges. Meanwhile, his diet consisted largely of Carlton Draught, sausage rolls and psychedelics. While some may have taken this to be a normal Aussie man on an odyssey of self-discovery, the truth is he was just impressively unhinged. One drizzly winter morning, as he and workmates were stopping by on the way to a job, he appeared through the doors and stretched like a cat that had eaten its fill, exclaiming: 'It's mushroom season, boys!'

Hugo's problem was that fishing in the dark wellspring of one's inner mind on psychedelics is, as a rule, wildly unpredictable. When you add in the fact that Hugo was already as mad as as a frog in a sock – even without any psychodelic assistance – you start to get the picture. You see, for some, psychedelics can open a shining cat flap in the door of reality. With Hugo, they ripped the door off its hinges, drove a car through the wall and then torched the house, leaving him cowering in the smouldering remains at the mercy of The Void.

One night, as I was stocking the magazines and finishing off a king-sized peppermint Aero bar, I heard the familiar cries of lunacy from down the road, growing louder as they approached. After several minutes, Hugo appeared at the doors, hammering with his fists to be let in.

He entered and turned to face me, his eyes like dancing green campfires.

'If the world ended tonight, would you be able to face your creator? Would you accept your fate? Or would you fight?'

'I'd have to think about it,' I said.

'I've *been* thinking about it. And you know what? The world has teeth, man. The world is like a shark, and it can bite you *whenever the fuck it wants*! We can't just lie down. Death stalks us constantly. He stalks *me*. He watches . . . Always. Always watching . . . I can see him now.'

At this point, a car appeared, its high beams bleaching away the black of the night. As I squinted, Hugo's eyes seemed to drink the light in. He let it fill them as a smile slithered across his sweaty face.

'He is here. The destroyer of worlds.' He bolted to the back of the store and yanked open the freezer doors, jumping in to hide behind a wall of ice.

The car which pulled up was actually Hugo's, and his mate appeared at the door.

'He's had two tabs of acid, some mushies and a bit of speed. He thinks he's dying this time. Oh, and that I'm death, and have come for him, fucking idiot.'

I pointed to the freezer.

Hugo's mate yanked open the door and grabbed him, wrestling him to the ground as Hugo shrieked. As my view of the fun was blocked, against my better judgement I left the console to find Hugo curled up into a foetal ball and whimpering up at his tormenter. He was shivering.

'I'm cold. So cold.'

'You've been in a fucking freezer, you idiot.'

'Life is leaving my body. Death, you will have your way.'

Hugo slowed his breathing, grunted and was still. His mate picked up a copy of *Penthouse* and waited. Stuff like this happened a fair bit with Hugo.

Okay, it's time for what what we've all been waiting for: gratuitous servo sex!

I had a few cute regular customers that would always pop in on a Friday or Saturday night, on their way out clubbing, to a house party or on their way home, and always they'd burst through the auto-doors and cry, 'You again! I always see you in here wasting your life away!'

They'd come in again the following week, staying to chat for a bit, remarking offhandedly that it must get lonely in here, with more sexual tension in the store than Frodo and Sam's journey into Mordor. It took me a while to catch on. And who could blame me? It doesn't matter how much porn you watch with similar storylines, the idea that it could actually happen to you isn't something anywhere near approaching conceivable at first. But when a short curvy brunette walked in still tipsy and asked to see what the storeroom looked like, all my retail Christmases had suddenly come at once.

And it was fun. Sex anywhere it's not supposed to happen usually is. As I soon discovered, there weren't too many conventional places available, but with a bit of creativity, I could figure out something. While the back office made the most sense, inside the fridge was definitely something different. I sort of wanted to try the freezer but was reticent for dare I say obvious reasons.

The most annoying part was those pesky customers that didn't seem to realise I was having a lot more interesting time pumping something other than petrol. It was the only time I'd completely ignore the beep of the pump authorisation system and take the batteries out of the doorbell. I felt I'd at least earned that right.

But, if I'm completely honest, stuff like that only happened rarely. In reality, there's not much chance of getting some while on the job. See, usually you aim for the women you work with. The problem was, most of my female co-workers were over twice my size, married and the wrong side of forty-five. I never really had the opportunity for a whirlwind work romance.

Well, there was this one time.

After stopping at another of my company's servos one afternoon, I fell into the eyes of a young girl with a perky smile. 'JACQUI', as her name tag read, was hot, even in a company shirt that was a size too big and tried unsuccessfully to hide her shapely curves.

I wasn't alone in my admiration: several guys were trying to drag out their transactions for as long as possible. I sneered at them, mainly because I was doing the same thing, then played my trump card. I told her we had the same job and we chatted for several minutes about stupid customers while the stupid customers themselves grew angrier by the second. I floated back to my car, amazed that a young sexy woman was actually under the employ of my company and that one day our paths might cross, and we could get to know each other a little better.

I didn't have to wait very long.

While receiving a stock delivery the following Friday night, I asked the driver (a massive, menacing block of a man – the same guy who smashed my door) as he was dragging a pallet of stock into my store, if he happened to deliver to Sunshine, say, on a Thursday night? Forgetting the massive pallet and releasing it where it stood, blocking access to the store, he turned and looked at me quizzically, his smile widening second by second.

'Jacqui.'

'Yeah,' I said, matching his grin. 'How'd you know?'

'Because I'm in proud possession of a penis, my friend,' he said. 'As are all the others that spend half their nights in there, seriously skewing the ratio of male/female customers every Wednesday and Thursday night . . . I'll put in a good word for you,' he winked, turning around to laugh at the mass of impatient but wary people waiting to make their way into the store.

He returned the following Friday with some promising news: 'She tells me she'll be in your neck of the woods next Saturday night and might pop in, so make yourself look pretty.'

Saturday came around and, on two hours sleep that night before work, I woke to an angry alarm clock at 00:21 am. Although I started at midnight, I wasn't fussed, as the bastard on before me was always late whenever I was waiting to finish in the morning, so revenge would be both sweet and productive, as I spent the following twenty minutes sprucing myself up for Jacqui's visit later that night.

I pulled into work just before 1 am, with cool messy hair and a faint miasma of expensive aftershave, and stared uneasily at the purple Daihatsu with a bright orange sticker yelling: THIS BITCH BITES! As I walked toward the store my heart sank as I realised who was standing behind the console. Turns out Jacqui was covering a shift there. She blasted me for a full minute as I stood there in shock, my dreams of a work romance in tatters. Turns out she had another job as a door bitch at a rave club in the city, which was now under threat thanks to me. With a huff she stomped out of the store and into her little car, speeding off and out of my life forever.

CHAPTER 16

DAYSLEEPER

There's a reason they call it the graveyard shift. Start working nights and everything you know will begin to wither and die: your health, your sleeping patterns, your appetite and any semblance of a social life. They will become tombstones, all, mocking you as you wander the gaudy aisles, stocking them with a never-ending supply of dog food, Tic Tacs, Kettle chips and king-sized Mars Bars.

If you think hard enough you can almost remember what it felt like to be normal. When you'd awaken to a chorus of birdsong and the morning sun streaming in through your window, instead of trying to block it out with a dark-coloured sheet so you could attempt the cruel folly of diurnal sleep.

I'm not religious, but daytime sleep is loosely what I imagine hell to be like. After scrutinising Dante's Seven Circles, I place it somewhere in the middle. Sleeping is hard enough for most of us already, but attempting it in the middle of a hot summer

weekend when the carefree masses are at play borders on impossible. The rasping sounds of suburbia scraped into my synapses
one by one as I lay there with homicidal thoughts growing ever
closer to reality.

There's the spluttering cough, then roar of a chorus of
lawnmowers, each more ancient than the one before it; the sharp
thock of cricket balls against fences; the shrieking of summer
water-fights; lonely housewives pumping Neil Diamond through
their empty hallways; the yapp-yapp-yapping of psychotic Jack
Russells; the screaming of whipper-snippers; and the revving of
every engine in existence as they scream open-windowed past
my house with their grating autotuned Top 40isms.

To compound this are noises from inside: the *whump* then
keening drone of the flushing toilet; the washing machine
attempting take-off; your three-year-old nephew addicted to
120 decibel Pixar films; and the shrill, insistent beeping of every
appliance in the house.

Earplugs, headphones, they help slightly, but when seeking
silken silence as a doorway to what you pray will be nourishing
oblivion, any noise is amplified into unadulterated discordia.
So many times, I planned exactly *how* I would demolish each
of these machines in an orgy of sweet violent destruction. As
my mind began to wander to places it shouldn't, my thoughts
returned to The Beep, the undisputed King of Discordia, waiting
impatiently for me, now only several hours away. I continued
to toss and turn, praying through gritted teeth for peace but
knowing it was now unreachable for another day. Instead,
I received only detritus from a world that had woken up and
left me behind.

I used to spend the evenings immersed in a good book, sighing comfortably in the anticipation of restful sleep as the rain thundered down outside, instead of lying in my bed late on a summer night as the shouts of a nearby party drifted through my window with the searing heat and humidity that had kept me company while I'd fitfully slept during the day. Bathed in my own sweat, I'd wake up a kilogram lighter, and smelling like I'd been baked in the filthy heat of Satan's arsehole.

You see, the body isn't meant to stay up all night and sleep all day. Shocking, I know. Ask most bartenders, bakers, taxi drivers and drug dealers how their body clock's going. Mine – like Salvador Dali's surrealist version – melted away long ago, liquefied by the heat of a Melbourne summer.

But even winters messed with my head. I'd get home from work at 6 am and sometimes sleep almost straight away, with the world still dark outside. I'd then wake in the early evening, having overslept, when, of course, it's dark once again. As a result, I got almost no natural light for weeks at a time.

This endless disruption of your circadian rhythm, not to mention the horrific lack of vitamin D, serotonin, dopamine and other goodies means, as the research continues to show, that working graveyards is linked with the following delightful maladies: cluster headaches, diabetes, depression, ischaemic heart disease and cancer.

Like a narcoleptic, I began dropping everything for precious, precious sleep. I downloaded forest sounds, rainstorms and whale noises, and played them through a state-of-the-art 5.1 surround sound system. I bought ineffective black-out blinds then looked into window shutters. I cornered the market on high-strength melatonin. I even turned to the Divine, buying a statue of Hypnos, the Greek God of Sleep, and paired him with his son, Morpheus, the God of Dreams, placing them on opposite bedside tables and desperately praying to them to spirit me across the River Styx and into the glorious quiet Underworld as I tossed and turned in my room's nacreous, sweaty gloom.

Waking up – something I was never particularly skilful at – became more and more of a struggle. You feel so much *crustier* when you've slept in fits and starts all day. All the comings and goings have come and gone, and you're standing there draining a litre bottle of warm water in the darkness, wondering where and what the hell you are.

But I had some help in that regard. Like me, my cat Moby was a Daysleeper, and a tried and trusted routine as my personal alarm clock. Like his namesake, everyone's favourite short, bald purveyor of downtempo ambient electronica, Moby was exceedingly chill. He slept on my bed curled alongside the crook of my neck, so I could feel him purring like a chainsaw on rocket fuel. Halfway through the day he'd stretch, as if he was about to whisper a secret in my ear. Sometimes his paw would stay there until the late evening, rested lovingly on my shoulder, my best pal. When wanting food or simple escape, he'd pad up to my pillow and simply sit on my face.

As I rose rapidly into wakefulness, I'd panic at the unseen force filling my mouth with fur. As I pushed him off, he'd offer a triumphant meow, as I leapt from bed to turf him out the door, cursing.

Once I'd dealt with Moby, I'd stumble into the shower, my car, and then work, in a sour hypnagogic daze. Drunk on heat, dehydration and lack of sleep, I was so hungry I could've eaten the arse out of a rag doll through the bottom of a cane chair, and often did, filling my poor guts with greasy chemical-laden roast chicken rolls and imitation cheeseburgers.

'Maaaan, you look stoned,' the stoners would tell me as they giggled up and down the aisles, always grabbing the Chicken Twisties, Mars Pods and Sour Cream & Onion Pringles before asking me if I had any more weed. Looking at my reflection in the mirror, I saw my puffy eyelids and the darkened bags below, like storm clouds massing in the winter sky. I was usually able to grab some coffee and put it out of my mind. At least until I saw The Lifers.

See, in the course of this job I've had every projectile you can fathom thrown at me, I've had my life threatened more times than you've had hot dinners, and I've seen a morbidly obese man wearing nothing but a loosely fitted nappy, fake nipples and a green afro. But even after all this, I never once thought it was time to pack it all in. Only one thing ever made me think twice: The Lifers.

They were washed-out old men, scaly creatures of night with forty years of factory shift work carved into lizard faces that looked more like leather handbags than any human countenance. Every wrinkle was sliced in with a knife, but with no blood evident, so dry were their ophidian forms.

They'd weave into the store, pick up a *Herald Sun* and warm their snake blood under the fluorescents. But it was their eyes that got me: two reptilian pissholes in the snow that looked inside me, feeding on my fear as they pulsed with some malevolent intellect. 'Soon,' they seemed to say. 'Soon buddy, soon champ, you'll look like this too. And then you'll be one of us. One of us. One of ussss . . .'

While it shocked me, as time crept by, I suppose I ended up caring less about my appearance. I think most graveyarders do in the end. You get used to certain parts of your week being a blur, and to people saying (in the nicest possible way, of course) that you look like death warmed up. By this stage, though, I wasn't overly fussed. My narcotic exploits were slowing down considerably, having lost their fine edge of enchantment. With no uni, my world was one of cimmerian, purgatorial gloom; squinting under fluorescents, trying to sleep and saving my pennies away.

Weekday mornings held this humorous kind of symmetry: the comatose tradies trudged into the store, blinking as they stumbled toward the Red Bull and V, while behind the console, I was fading, sipping bad coffee, both of us smacking into different walls of the same cruel obstacle course of sleep deprivation.

We understood each other, through our muffled yawns and bleary minds, and we didn't push things, because we didn't have the ability to. At 5.15 am, no one cares about discount dockets and cents per litre. They simply want sugar and caffeine in their bloodstream or they want to go home to bed. Until they get it, they are easy zombies, which suits everyone just fine.

It's those damn early-morning neurotic joggers that get to you. You know the kind: the early-middle-agers that have decided to *do something* about their fitness and go power-walking up and down the pavement, flushed with a healthy glow and brisk pride. They bound in and grab their paper, marching toward the counter with big whooshing breaths and iPod minis strapped to their arms playing uplifting baby-boomer tunes. Oh, how I hated them.

Daywalkers will never understand Daysleepers. We're just too different. They bustle about all day, replete with sunshine and normalcy; we spend our hours stocking shelves, pouring beers or herding freaks in neon-lit blackness. We graveyarders, all quarter-million of us, actually take the time to converse and interact with the world. There's no rush, you see. People are different at night, for night is when we play. Dance. Drink. Fight. Fuck. As the sky darkens to velvet, then black, we take risks we normally wouldn't and remove our masks, buoyed by the freedom of the stars.

Fifteen-second transactions can turn into hour-long conversations on everything from your footy team's chances in the finals to exploring the various schools of thought regarding existentialism. Some nights I was sage-like, imparting my wisdom into young, eager ears, while on others I was the student, consuming obscure facts and tales from all the amazing people that only come out at night. I've chaired fiery conferences on gender relations and played chaotic drunken games of two-a-side soccer, with the storeroom door and auto-door as opposing goals. It all depends on the person.

Daysleeping also gives you impressive brain fog. This never-ending jetlag takes your focus, concentration, short-term memory and any problem-solving ability, and pours a thick, cloying molasses over all of them, often making you step back and marvel at how you are, night after night, entrusted with control of a multimillion-dollar chemical deathtrap.

Graveyarders are simply unparalleled in their ability to lose things. Whether it is shift reports, staplers, money, their temper, a pair of scissors or simply their mind, none of us end the night with what we had when we started it. Pens are my thing. The one I'm writing with right now on this register receipt roll will be gone before the sun is up, beginning its journey to the same place those odd socks disappear to through that secret tunnel in your washing machine.

Barb could never figure out what I did with them. One morning as she was looking, to no avail, for something to write with, she grabbed me in a headlock and squawked, 'You've eaten them, haven't you?' as she scanned my mouth for telltale signs of ink.

I've lost count of the times I've actually forgotten how much change to give the customer even though I saw the number flash on the screen less than three seconds earlier. Some mornings you look over your shift data, marvelling at astronomic variances you can't even begin to explain. It's like part of your night was spent underwater, watching the bright little fishies float by.

So, there I was, kicked out of uni, with nothing on the horizon but fluorescent downlights, sleep deprivation and whingeing angry customers. My Faustian-pact with narcotics had swapped to vitamins and supplements as I futilely tried to fix an internal injury by sticking band-aids to myself. My mind was a foggy airport terminal, where a thought – like a piece of lost luggage – would slip through the black flaps of my subconscious and onto the dusty baggage carousel of my mind. But with so much fog in the terminal I needed a spotlight to find it. I couldn't even *read* properly, let alone write.

It was mid-2004 and everything was becoming grey. A dull, lifeless grey. It was like in the days following the comedown from a pill, but this time it seemed to be perpetual. The customers trundled in like lemmings on purgatory's conveyer belt and my heart sank, as I wondered, not for the first time, if this was what I was going to be doing five years from now. Ten years from now. I didn't get it. I was a *smart kid*. I breezed through school. I wasn't one of the fuck-ups . . . But maybe I *was* now. Maybe working in servos was where I belonged?

As the nights went on, I sank deeper. A jaded nihilist on auto-pilot, trying to block out the customers as my mind drifted to places it shouldn't. I had nightly fantasies of the whole servo immolating me in a searing ball of orange flame, mushrooming up into the suburban skies.

I'd got to the point where nothing made much difference anymore. All the days and nights joined like one long road leading out into the neon wilderness.

I woke one night, already late for work, and didn't care whether I got up or not. The clock's buzzing had a demonic quality about it, a metallic braying that grinded into my brain. Even its numbers looked evil; searing scarlet slashed through obsidian blackness, but with such callous precision. Some force had gathered while I was sleeping, and its edges had been busy, filing themselves to wicked points. This force wanted me to suffer. I tried to shake off sleep and crawled out of bed but felt only a dull thud bouncing around my skull as I trudged toward the bathroom. My running nose showed the early signs of another cold and my muscles ached with lack of sleep. Realising how cold it was, I made for the shower. The hot water had run out. Enough.

I rang my retail area manager and announced I was off to Europe. While at first, he didn't like the cut of my jib, me threatening to quit soon changed his tune. Reliable weekend graveyarders are solid rolled gold and he knew it in his bones. Checkmate, and three glorious months off.

Funnily enough, in the last week before leaving, my depression seemed to melt away. I felt like a groggy bear padding out from my cave, stretching and blinking in the glorious spring morning, seeing the distant river and its bountiful salmon leaping in the sun.

I finished my last shift and hours later I was at the buzzing hive of Melbourne Airport's international departures terminal, suddenly realising that all the shit that enervated me – the dickheads, the freaks and the maddening, fatalistic *sameness* of my hometown – none of these things were allowed to board the plane with me.

Beginning to notice the shafts of sunlight now pouring into the chrysalis that had contained me for the past fifteen months, I strode up to the smiling dollish lady at the counter, her sing-song voice wishing me a pleasant flight as I handed her my boarding pass, joy blooming fiercely in my guts.

And down that fateful tunnel I tramped, the corners of my mouth pulled inexorably toward my ears until I resembled a madman – a dangerous smile in an airport – as I passed another beaming ingénue before strapping myself in, thrumming like an arrow in a bow.

And I speared down that runway, soaring up to meet the clouds, with everything that used to define me now earthbound. All the same miscreants and automatons would trundle in tonight, whingeing about the price of smokes and petrol, but I'd be at cruising altitude, on my way to Neverland, pushed by two mammoth Boeing engines, glinting in the newfound sun.

My mate Oliver and I met in Singapore and landed in Lisbon, smack-bang in the middle of the Euros, the second biggest soccer tournament on earth. The entire country was buzzing in a football-induced frenzy and as we stepped into our hire car outside the airport, the streets were full of drunken Portuguese celebrating Portugal's 1–0 victory over despised big brother Spain. As we battled through the streets, trying to stay on the right side of the road while looking for the freeway, swarthy whooping men hung out of car windows, clutching bottles of wine and Portuguese flags that whipped from their hands in the stiff breeze.

After nearly being killed at several roundabouts, we finally made it onto the freeway, on which all the traffic moved faster than shit through a goose. Although we pushed our little Fiat Punto until it shook, rounding 140 km/h, we were constantly passed on our left by a blur of Ferraris, Mercs, Beamers and Porsches.

We pulled into a massive servo, linked to the freeway by a long winding exit ramp that ended in a vast city of pumps, all of them shining and new, filling shiny new cars, driven by well-groomed, fresh-faced Portuguese.

As I walked in to pay, I checked out all the foreign chocolate bars and bright fruit drinks, longing for a sausage roll and a Chocolate Big M, as I compared the store to mine back home. I approached the console and marvelled at the shiny, happy people behind it, all smiles and buttoned-down collars, their ties sharply knotted and pearly white teeth on permanent display for anyone who crossed their vision. Armies of workers strolled the aisles, straightening the placement of already-straight products, chatting excitedly to anyone within a 10-metre radius.

Customer after customer filed in cheerily and exchanged jovial, rapid-fire conversation with the attendants before efficiently swiping their shiny card or extracting their crisp euros without a single grimace or frosty glare, even though the petrol prices greatly exceeded ours back home. When it was my turn, after much pointing and waving, I managed to pay for our pump and was ready to walk off when I uttered what I naively thought was 'thank you' in Portuguese: 'Gracias'. The colour drained from the console operator's face like someone had opened a plughole as he stared me down, before scolding me with a, 'No España! Nós dizemos O B R I G A D O. Obrigado! No España!' Feeling like I had been sent to my room without dinner, I walked the gauntlet of accusing stares back to the car.

As we flew down the freeway toward Faro in the south – at first hugging the coastline, with its red cliffs shining beautiful against the blue of the North Atlantic, and then sailing inland over olive green hills lined with verdant orchids – I grinned stupidly to myself, not just because I was in a beautiful foreign country, but because I was up and awake during the day – and free, with no work that night, or for several months – my head stuck out the window like a giddy dog, savouring the rich smells from the breeze slapping at my face as I bathed in the sweet glow of the Portuguese sun. I thought of Claudia in that moment, the memory of her sprawled blissfully across the grass. I grinned wider. I'd woken up. And it was fucking glorious.

We soon made it into the baked Algarve region of the south, and on to Faro, as Holland knocked Sweden out of the European Cup on penalties in a half-finished behemoth of a stadium, filled with equal parts yellow and orange; teeming masses of colour, one elated, one destroyed, both dancing and singing for over 120 minutes as the scorched afternoon became night.

We then went north, to Porto, the UNESCO city of bridges, where we stayed at a beautiful hostel that overlooked the merging of the Atlantic with the mighty Duoro River. Porto is stunning. A city that feeds off its colossal river, banks that climb quickly skyward, with centuries-old forts, castles and crumbling grey chapels. White limestone houses were hewn out of crumbling cliffs. The lush green of trees merged everywhere with the thousands of red roofs that baked unhurriedly in the sun, the very colours of Portugal's flag.

Every night was a party, as DJs mixed cocktails and records at tiny dance parties for beautiful people on the golden sand, replays of the day's soccer matches on big screens everywhere you looked. The city was a friendly raging torrent of nationalities, all drunken chants and cross-cultural diplomacy. I headed across the river to a rave one night in a massive park with a stunning backdrop of a seventeenth-century haunted-looking mansion, with holes like Swiss cheese, lit up by dozens of lasers that scythed through the night air as fifteen-year-olds came up offering handfuls of pills, at 2 euro each.

When the Greeks beat the Czechs with literally the last kick of the match to send their unfancied country, who had never come close to winning anything, into the final, the stadium burst into instant chaotic Nirvana, as we were mobbed by dozens of neighbouring Greeks, tears streaming down their faces as they wailed in delirious joy, reduced to babies by the sheer realisation of their outrageous dreams. As we partied long into the night, every second Greek supporter we met was from bloody Melbourne.

Portugal then beat the Netherlands to set up a final against Greece – a baby-faced Cristiano Ronaldo just starting his career – and the entire country exploded into joyous bedlam, giddy crowds rocking, overturning and jumping on the undercarriages of unfortunate cars, every window in the city hanging a Portuguese flag that week.

I blazed a pretty stock-standard trail through Europe for a first-timer, Contiki-ing, running from the bulls in Spain, being treated rudely and impatiently in Paris, salivating over steel, food and flesh in Italy. The works. There were so many places of just silly, silly beauty. One place I stayed in France was arguably the best of them all. It was a chateau just north of Lyon in France's Beaujolais region and it was seventeen kinds of gorgeous. An outdoor double staircase led down to an endless sea of perfect green grass, broken up only by the cerulean of an in-ground pool. Lying on that flawless green carpet, with a litre bottle of Beaujolais as the sun sank behind the patchwork hills of the surrounding vineyards, I tried to remember what work was like back home, but for that blissful second, as I stretched like a lazy cat in the warm Gallic sun, I actually couldn't.

On a night out on my Contiki tour, after a few 1-litre steins in a beer hall, I dropped into a servo in Munich where I introduced myself to the console operator and told him we were kindred spirits; that I was a console operator all the way from Melbourne, Australia. At this he snorted a sarcastic laugh that only the Germans seem to have perfected. 'Yes, very good. I don't care. I hate this place. I'm actually quitting in the morning. You can take what you want. I will decide the price.'

'Are you serious?'

'Why not? I'm quitting, so what would I care?'

I instantly started grabbing everything I could fit in my arms before he changed his mind. He then totally floored me by offering me a box. I brought it all to the counter and he glanced at it and said, 'Ten euro.' I gleefully handed over the money and left with a massive box of crap worth well over 300 euro that supplied the confectionary needs of my entire tour bus for the next week. Part of my haul was three bottles of Dr Pepper, which I used to woo the hot Texan girl in our group who lived on the stuff. No dice, unfortunately. She was the religious type.

More than anything, though, I rejoiced in waking up every *day* and actually going to sleep at *night*. It sounds stupid but after being a vampire for so long, it was revolutionary. I ate breakfast in the morning, lunch at lunchtime, and dinner in the evening. Not one microwaved roast chicken roll or bottle of V among them. After only three weeks of a European summer, I had a proper tan, and stopped looking like death warmed up.

I remember sitting on a stone bench, in a lookout park, gazing out at the Amalfi coastline. I stared at the stone cliffs that reared up like startled horses from the sparkling blue of the Tyrrhenian Sea, at the apricot-coloured mansions basted in falling sun, full of beautiful people who had never worked a nightshift in their life; never been pinned down by the eyes of a speed freak at 3 am while the pump authorisation system ground into their brain.

And then it was over. I caught the train up to Rome and, due to running severely low on funds, spent the night at Leonardo Da Vinci airport, where all sorts of misfits and grifters shuffled around the empty terminal, stealing anything that wasn't tied down. It was like being back at work.

The following day I flew home and returned to the night. Like a dog to its fresh spew. After thirty sleepless hours on three planes and a taxi home, I showered and crawled into bed, turning on my mobile, just in time for it to ring, with my retail area manager on the line.

'Hello Dave, how was your trip?'

'Ummm, yeah, great.'

'Any chance you can start tonight?'

'Innn . . . four hours?'

'That's the one.'

I walked through the parting glass to see Nellie, standing behind the console and leafing through some hardcore porn.

'I knew we did breasts but I didn't know we did crotches too!'

She was halfway through serving a sheepish-looking middle-aged man now regretting his choice of servo for some in-and-out stealth porn. He had the money ready in his hands and was looking pleadingly at her, but Nellie hadn't finished perusing.

'I can't believe we actually do crotches. Oh well, enjoy your crotches, sir.'

The guy thrust the money at her, snatched back his change and magazine, and stomped out of the store.

I walked up to the locked console door and laughed, having been away so long I'd forgotten the code. Nellie let me in and smiled warmly.

'Hello David. How was Europe? And did you know we did crotches?'

'Hello Nellie. Europe was magic. And yes, we've always done crotches.'

'Oh okay. Did *you* do any crotches in Europe?'

'A couple, yeah.'

'Good boy. Did you go to Yugoslavia? No? Next time, go to Yugoslavia. Do you remember what to do here? You've been gone so long.'

'I should be right.'

'Okay, you count up your till and I'm going to have a smoke. I've been waiting two fucking hours to have a smoke.'

'Go home, Nellie, I can manage.'

'It's good to have you back, David.'

'It's good to be back.'

'No it's not. You'll see.'

CHAPTER 17

THE WOKEN DREAM OF EVERYTHING

And I did see. The paleness and dark hateful circles around my eyes soon greeted me in the mirror like old friends. After bathing in exalted freedom and sun-kissed joy for months on end, my return to nights hit even harder. Standing behind the console and waiting for underage kids to realise they weren't getting their smokes or for the wheels in the addicts' heads to turn and deduce they had no money in their accounts for smack, I noticed that time had seemed to slow down.

I willed every passing second on, trying to speed it up with the sheer power of my mind. After putting away the milk delivery, stacking the newspapers and stocking the chocolate bars, I'd return to the console to see almost the same glowing green numbers as before. A job that I was certain took half an hour actually took seven minutes. I desperately wanted the peace that I'd felt in Europe back. But with the Mediterranean sun replaced by cruel fluorescents and the worldly Europeans replaced by whingeing bogans, I found it ungraspable.

No: I found it infuriating. Ignoring for a second how blissfully lucky I was to have won the geological lottery, being born into a free, beautiful and affluent country that was a shining oasis compared to two-thirds of the world, in my delusion I felt I'd been dealt a needlessly cruel hand. How could such ease and beauty exist when I was stuck here under the sour lights herding addled vindictive fools? Every night I felt myself acquiescing more and more to Andrzej's bitter and saturnine ideology of war. But this time, the anger felt deeper, more a part of me than a separate idea. I needed to come to terms with who, what and where I was. Or I just needed to detach from it somehow.

Work any maddeningly boring job for long enough and you learn fairly quickly that there's one skill that, while you won't pick it up in training, happens to be the most important of all. In short, you need to learn how to uncouple from reality. How to flick the auto-pilot switch and become *not really there*. Secure in your own world, everything now bounces off you: the lights, the shouts, the stupidity and the impressively belligerent regard in which almost everyone holds you. It's your special place.

I wanted my nights to flow by like a river, while I lay deep in the middle, coursing with the gentle tide. I didn't know how this would happen, exactly, but I knew I needed it like a dying houseplant needed a bucket of fresh rainwater. And I knew that if I didn't get it, I would soon go so postal that not a single scintilla of normal everyday life would ever be able to reach me again.

So, once more, I turned to the Divine. I started buying spiritual books, learning deeply about mindfulness and the unending mysteries of human consciousness. I read Eckhart Tolle's *The Power of Now* and the clarity of its simple message rang in my head like a shining bell: the past is gone forever, while the future *never actually comes*, because it is *always now*. Several times while reading, I actually had to put the damn book down: its razor-sharp prose and glinting rationale cored me like a soft ripe pear.

That book rapidly became my drug. Or, at least, an effective volume knob, for as I read it, all external thoughts drifted away. All my regrets, anger and shame of the past, and equally, all my lingering anxieties of what was to come tomorrow, in a few weeks, even fifteen minutes from now. Then I started worrying about how long it would last for and, *poof,* all the ugly noise came back. But I kept reading and back the glorious hush would come. Instead of detaching, I realised that all I really needed to do was immerse myself fully in the moment as much as possible, to surrender to it, and allow its all-encompassing peace to swim through me.

But my ego, not interested one iota in all-encompassing peace, kept yanking the reins from my pious hands. *Think*, he entreated with an oily smile, my serpent in the garden, fatly coiled around my cortex. *It's what your brilliant writer brain does best, after all, Dave. You have to think about the past, otherwise, how would you write about it? And the glorious future! Who knows what treasures could come! You can solve all your problems with the power of your mind.*

Extending my middle finger toward that scheming prick, I resolved to triumph over him. But try as I might to stay present, some random thought thrown up from the deep valleys of my subconscious would lead to another and, before I knew it, I was back, once again shackled with regret from the past and resentment about what new freak would come barging in through the parting glass.

So, on the advice of Mr Tolle, and the jovial Rathindra, I learnt to meditate; to quieten my mind, that halogen-lit racecourse of ghouls, and connect to a peace I was utterly ravenous for.

Outside of booking a Vipassana retreat, I had no idea what to do, no concept of how to reach any kind of stillness, but I was committed.

I soon found that there were many forms of meditation, and many ways to meditate. While some of you may thumb your nose at the idea, everyone does it already without noticing. Meditating is, essentially, achieving an inner stillness by not reacting to your thoughts, and thus giving your brain a blessed lunch break. It took a while for me to realise the utter simplicity of this, but once I did, I was able to plug into what lay beyond my mind. But not without a fight.

My ego had other ideas. Shocked at the sudden snub, he emptied the arsenal of my subconscious, flinging any thought, memory or emotion he could at me, hoping desperately that something stuck to my walls. But I kept on, softly resolute.

It went kinda like this: imagine yourself in a messy, darkened room. It's littered with month-old cheeseburger wrappers, dirty socks, ripped magazines and half-crushed cans of Red Bull. In the middle of this mess sits you, your attention stolen by the countless neon posters that flash up on your darkened walls. Constantly appearing, for different lengths of time and in varying degrees of brightness, all soon vanish, but are immediately replaced by others, so that there's never a moment of complete dark. Or peace.

These posters are of course your thoughts; your hopes, worries, doubts, resentments, fears and insecurities. They yell in searing, lurid iridescence, shouting that your arse is too big; that your cock is too small; that this week's electricity bill is due; that the milk in your fridge is off; that you're failing a worthless uni degree; that you're forty-three and a lazy fat slob that will never amount to anything; and on they go . . . These are the mechanisms of your mind.

You whip your head from projection to projection, lurching from guilt to joy; from annoyance to hope; from wonder to self-loathing. You're stuck in Plato's Cave, a prison so dark and dependent on shouty neon that you can't even see the glorious sun that shines just outside the blackened walls that hold you within.

Imagine for a second that if you ignored all that fluorescent bullshit for long enough, that seedy cinema fades away completely and, now seeing it for the desperate dancing illusion that it indubitably is, you escape to the surface, out into a soft blue sky and green fields, rustling in a warm breeze underneath a brilliant shining sun. And in that soft warm field is where you feel the pure white light of what you actually Are. *That* is meditation. Eventually. If you stick to it. Sorry to get all Eckhart on you, but that's what it was like for me.

I started by looking at the slurpee machines. There's something hypnotically Zen about frozen water churning soundlessly in brilliant colours. Yellows, blues, reds and greens filled the plastic barrels as they roiled endlessly, as cold and devoid of life as the vacuum of space itself. And yet, they *did* have life:

Dr Masaru Emoto had shown that water indeed held consciousness, altering its molecules to match the energy directed at it.
There were answers to mysteries I couldn't yet fathom within
those hypnotic icy swirls. And I realised then that, being 60 per
cent water, those same answers swam in me too.

So, in the tail-end of a fierce Melbourne winter, in the long
stretches between customers, I stared endlessly into the gyrations
of that steel churner as its tip emerged from the icy sea of
colour, only to turn slowly under the surface again, hidden by
the wonder of that sweet frosty goo. And as I did, harsh reality
fell away into a soft humming realm of light.

When I rejoined the world, I discovered a wonderful thing:
it no longer enervated me. The shouts of deranged customers,
the screeches of tyres and the autotuned dissonance from thud-
ding sound systems had all become background noise; a simple
transfer of aural energy, neither good or bad. It just *was*.

Suddenly, I no longer attached any importance to the events
that used to drain me so much. What mattered now was peace,
wisdom and enlightenment. My brain's lunch breaks were getting
longer and more numerous. I felt sharper, lighter, more lucid,
but calmer at the same time. My other world had converged
with this one and it was more beautiful than I ever thought it
could be. A supreme stillness lingered now, and it was everywhere. Although nights can be an incredibly desolate time at
a service station, I began to finally fall in love with their dark
secret beauty.

The orange streetlights burned off endlessly into the night,
casting their lusty breath onto the asphalt below as moths came
from everywhere to flutter in the electric glow of their addiction.
I strolled the empty forecourt and a lone seagull followed. The

only sound was the soft drone of approaching taxis and the sighing of their overused brake pads as they glided through the roundabout. My icy breath floated skyward, toward the impossible smear of stars. Something about their ridiculous vastness brought me even more peace, until the low rumble of a freight truck and the approaching glow of its headlights cut short my meditation as I headed back inside where the colours and heat wrapped around me like an excited hug from a hyperactive five-year-old.

There's a time during graveyards, usually 2.30 am to 4.30 am, when nothing of any substance or direction really occurs. People are neither starting nor finishing shift work. The coffee machine's contractions disappear and a kind of peace spreads itself among the aisles. I look out at the night and it's velvet, serene. The servo canopy lights are tableclothed in shifting ghosts of thick fog. Current affairs await me on the newsstands while sausage rolls beckon with their crusty, meaty warmth, spiced with sweet tangy tomato sauce. *This is why I work nights*, I'd think to myself as I settled down with a *TIME* magazine and a Chocolate Big M, watching the ants hold a silent requiem around their fallen, Morteined comrades. There were interruptions, but they comprised only trivial things: bread, milk, taxis, various emo freaks. For these two hours, I was at peace.

This peace slowly began to spread to all areas of my job – especially customer conflict. Meditation calmed me, because my ego was finally taking a back seat. In the past, when someone was ridiculously rude or belligerent, my ego would scream, *You're not gonna let them get away with talking like that, are you?* And I decided that, no, I wouldn't, so the red mist would inevitably descend.

With the benefit of distance and stillness, I now realised how futile the whole thing was; the whole cycle of 'I'm tougher than you and I can prove it', all to please our ravenous egos. Why bounce all this bad energy back and forth between us in a spiritual squash match? I could see now how deeply these people were at war with themselves, and how such a battle could never be won. Anyone can wage war, but I chose to break that torrid, exhausting cycle and spread peace instead.

The best part was that this peace was portable. I continued to meditate, replacing my slurpee machines at work with candles at home. I'd glide homeward in the charcoal murk of early spring mornings with psyambient music pouring softly through the car, losing myself in the wash of opposing headlights, a river of impatient gold that stretched to the freeway as they shunted and honked, headed for office cubicles and telephone headsets, their horns betraying their turbulent minds, foreshadowing the worry and stress of the day to come.

Lying in bed before work, I'd listen to the wind howling like a banshee as it sneered in at me through the silver panes, its rain spattering against the glass that rattled and creaked. Mirroring my peaceful state, Moby was reclining in the diamond eye of the tumult, in his current preferred sleeping position – flat on his back with his tail splayed out and all four paws sticking crookedly up in the air – like an overturned but contented cockroach.

Silvery trails from twin incense sticks twisted in the air like dancing dragons, entwining with each other as they pirouetted above the tunes leaking softly from my speakers. My room was a smoky Bedouin tent in the middle of suburbia, and I sat ensconced in the softest, most accessible magic I'd yet known. Filters that had long sat over my perception faded away and unfettered reality swam in to replace it. It had been less than two months, but meditation had already softened me; stilled the stuttering glories strobing broken-promised in the gaudy ballrooms of my skull; quenched the roaring furnace-dreams that had me hoisting lusty ladders to blazing narcotic suns.

I was the woken dream of Everything as I sat in lotus position in my cocoon of gentle acceptance, watching the red minutes blink on by and laughing at the absurd concept of time. Nothing mattered, and it was beautiful.

Back at work, with new eyes, I saw a new world. There was this gargantuan huntsman that took up residence above pump three. I saw him one night when I was strolling the empty fore-court at 3 am, stretched out across the bowser's display screen, bigger than my splayed hand. In the past, huge spiders always gave me a jolt, but staring at this one I actually felt connection and a respect for another living being. For several months he lived there, feasting on the maelstrom of moths that pooled around him like a soup. He'd sometimes crawl down the hose and straight onto the hands of unsuspecting customers, always women. He eventually met his fate under what I guess was a tyre. I went out one night to see him tattooed onto the forecourt.

There was one customer who seemed to typify the peaceful creep of the night. I named him Confucius. He worked at Toyota, although I'm not exactly sure what he could have done there. He came in once a week to fill up after he finished work and he was the slowest customer I've ever known. In my new state of Zen, though, I appreciated his glacial pace.

He had this warm, knowing air about him, and even though he moved with all the speed and deliberation of a sleep-deprived caterpillar, it was contagious. He'd pull up in his little Corolla, inching toward the pump like he was docking an ocean liner. I'd head off to stock the fridge and come out ten minutes later, just as he was slowly emerging from the car. He'd gaze at the pumps, regarding the different fuels on offer like metaphorical paths of chance: Unleaded could be good now, but Premium Unleaded may bring great prosperity in the future. After deciding on a pump and finally placing it in his car's outlet, he would pump it, 5 cents at a time, taking another ten minutes, while he watched the comings and goings of the pumps around him with a curious and cheerful detachment. Eventually finished, he'd begin his slow journey to the store, shuffling toward me as the cars waiting to leave the pumps would edge behind, revving impatiently for him to move before tearing off into the night.

Once he was inside, I was able to see him in all his stately magnificence. His Toyota shirt was too big for him, so my mind's eye saw a flowing robe that wasn't quite able to hide his prolific gut. His thick Coke-bottle glasses were topped by giant grey eyebrows that clung to his face like two mummified praying mantises. They curled outward and down, these prodigious brows, in long tufts from his round head, just like his name-sake, while his silver beard hung down to his gut like a vine.

He'd stand at the console and greet me merrily while looking down studiously at his wallet, considering which was the correct card. He'd talk about whatever came into his head while he did this, imparting small pieces of wisdom as he waved his card around in his hand, forgotten. The transaction finally over, he'd chuckle again, and keep talking slowly to me as the cross-armed line behind him grew more impatient with every word. Eventually, one of them would muscle in behind him, literally pushing him out of the way and slamming their card or cash down on the counter and turning to him with a prepared snarl. Confucius would merely chuckle once more, give me a fond farewell and start the long hike back to his car, shaking his head and chuckling as he went.

Like Confucius, I floated around the store, like wind inside a letterbox. I tried obscure chocolate bars, attempted origami and wrote meandering poetry, as my spirit began to smile. It was around then that my retail area manager called me to ask if I wanted to join civilisation and start working some day shifts. For some reason, I heard a disembodied voice emerge from my own mouth to say that I'd stick with nights; I felt at home.

I began to settle in, bringing my toothbrush in, then my razor. I'd just finish rinsing with mouthwash when there'd be an impatient pounding on the doors. Or I'd be all lathered up and ready to go when the doorbell would ring, with the Night RAM standing there, waiting patiently while I laughingly towelled myself off.

I brought in a toaster, an electric hotplate and some proper tea. I started eating real food, instead of the cancerous crap that surrounded me. I made curries, laksas and pancakes. I giggled to myself one night as I prepared chicken and salad sandwiches in the back room, slipping them into tamper-evident cash drop bags, ready for hungry nights to come, like some industrious housewife.

I sat there on Saturday nights, enveloped in my fuzzy warmth, sipping chai, eating organic sourdough spread with pesto and melted mozzarella, and reading *The Age* while the maelstrom of suburbia raged around me. Outside was Armageddon: wind howling like some demonic wolf, pitching rain sideways and spattering the grimacing faces of intrepid clubbers and surly taxi drivers as I sat inside sighing happily, a nocturnal monk finally at peace with the petty connivances of the world.

I kept a pet, too. Well, sort of. There was this decrepit white cat with one eye and what looked like three ears. Every morning at about 2 am he'd come mewling at the auto-doors and I'd write-off some meat to put into a bowl and watch him tear at it, like feeding time at the lion's cage. I named him Noctem, Latin for 'night'. He'd stalk the empty forecourt hunting insects, before a pair of approaching headlights sent him scattering into the safety of the bushes behind the car wash. He showed up at the door, once, silent, due to the gift he'd brought me: a massive bird-sized locust, twitching and jerking in its death throes.

As the nights pressed against the glass, I tried to answer the big questions. Like, how do all those bugs get inside the fluorescent lights? And if they can get in, why can't they get out? I think I spent two weeks on that one.

I loved watching the night slowly become day; the blackness fading to a charcoal as the trees across the road loomed as sentinels in the gloom, their shade finally darker than the brightening sky. I knew then that the tunnel had light; not just a pinprick, nor the leering, onrushing train of yesteryear, but a soft-growing radiance I felt in my cells, the same one that climbed up to feast slowly on the stars.

I served customers with a genuine smile then. Rathindra beamed at me like a proud parent as he witnessed my growth. Even the customers were taken aback.

'I must say, I've never received such compassionate and dedicated service,' one smiling woman offered at five o'clock one October morning, asking for a customer feedback form so she could tell my bosses of her positive experience.

'Well,' I offered, with my own wide smile, 'it *is* a service station after all.'

When I eventually finished my shift, the walk to my car was soaring joy. As I parted the electronic doors and strolled across the forecourt, the world stretched wide away from me like a slow-moving rubber band. Beams of early sun shot across the forecourt, splashing the bowsers in morning gold. The leaves of gums across the road became oily chandeliers. Orange clouds floated free in the giant blue bowl above as I marched to my car, beaming, and out into the great laugh of mankind once again.

This bliss held out for the better part of a year. Nothing could break it. Not Dubya being re-elected, not my sister nearly being wiped out by the Boxing Day Tsunami, not even The Black Eyed Peas' horrific 'My Humps'. I did nothing but work, read, meditate, write poetry, talk to Moby, and go on gargantuan walks by the local river every dusk, which were really just another form of meditation, all things considered.

Slowly though, the wheel of time turned, and out came drama, conflict and all-round total shitness.

It was mid-2005 and my mum and stepdad's marriage, having had a brief Indian summer, resumed its disintegration in earnest. My mum had a nervous breakdown and moved out, and so did I. I also had some major gut issues that seemed to spring up from nowhere then hang around permanently. Then I found out my dad had Parkinson's disease.

All the stillness and soft joy I'd lovingly built up shattered. It wasn't the best of times. Summer crept around again and day-sleeping was an ordeal that left me musty, sleep-deprived and losing weight, regardless of how much crap food, sugar and caffeine I once again shoved down my quivering cakehole. Do nights for long enough and they murder you, no matter how much drugs, vitamins or spirituality you throw at them.

At work, the noises of the store, so peaceful months ago, began to grate again. The dying fans of the deli fridge sounded like crickets on crack, while the coffee machine would try to give birth every fifteen minutes, just as I was starting to find a smidgeon of peace. Then there was the sporadic backfiring of cars, the increasingly terrible autotuned music sprayed from open windows (Fergie, it seems, did get to me after all) and, of course, the constant *BEEP BEEP BEEP BEEP* of the pump authorisation alarm that had ground into my brain and reappeared as a shrill stabby demon in my dreams.

And so 2005 leaked into a muggy, shouty and sullen 2006, and I soldiered on. Against my better judgement, I started another uni course (advertising – looking around at the pathetic ads plastered all over my servo by our inept marketing department I figured I could do much better). But already I was falling asleep in class again. I'd be listening to a lecture about integrated marketing communications and my head would droop forward until I snapped awake to the thinly disguised sniggers of my new classmates.

Beyond that, I worked, saved and behaved, sublimating my weariness, trauma and rage into a permanently sardonic smile and the months crawled by, then started to fly. Suddenly it was late 2006 and, skolling my third can of V for the night, I was standing in front of the mirror in the staff toilet I'd managed to block again (proud Stevo moment) with the detritus of too many stale pies and slimily toxic roast chicken rolls. I felt like shit on shiny porcelain that night, and my life was even worse. I was way behind on my uni assignments. Again. I wasn't writing. I wasn't meditating. I was simply standing on that worn anti-fatigue mat hurling sarcasm out into that kaleidoscopic void as the months fell by like rain. Or I was slumped on my computer chair at home, stalking MySpace profiles, reading pointless message boards, or wanking morosely to the internet's growing glut of accessible porn. Dad was getting noticeably worse, stiff legs struggling to walk at all, along with his ability to form proper words.

As I stared at the harrowing reflection in the work mirror, I ignored the shrill beep of the pumps and the pounding on the auto-doors, focusing instead on the dark piles of baggage under my dead pink eyes. I felt forgotten. Like I was trapped down a deep, musty well somewhere, looking up at that faint coin of light, where somewhere people played under a loving sun. Laughing. Dancing. With all the delights that came with freedom.

I knew I needed some glorious intervention. I needed fucking Neverland and I needed it now. The next day, I gently but firmly explained to Barb that if I was going to continue to be her crazy but dependable graveyarder then I would require quarterly – at the very least – weekends off to recharge. Where I could simply be, with no pump numbers, promo deals or psychotic weekend warriors pushing my buttons. She not only acquiesced, but moved heaven and earth in a poorly staffed store to make it a reality, whenever I needed it. Thank you, Barb.

I learnt quickly to treasure them, so that I not only had some memories for my mind to replay in gilded sepia tones as the psychopaths blustered in through parting glass, but also something to sustain me through the nights that followed, as I gazed greedily at that next pinprick of light, way off in the distance, but growing rapidly.

Looking back, they were simply dissociation from my troubles. But then, don't we all affix a pair of new glittering wings now and again to fly away from things we find disagreeable? I clung to those weekends like they were the last helicopter out of Saigon, spiriting me up and away from my fluorescent samsara and into singing clouds of joy.

I used them unapologetically for a myriad of fancy things. Hiking, paint-balling, wine-tasting, go-karting and mini raves in my lounge room with mates, spent in narcotic puddles of grinning bliss. Basically, anything that reminded me I was still alive and not always chained inside my bright little cage. I timed most of them for summer, especially music festivals.

CHAPTER 18

HALLMARK MOMENTS

El Diablo and I drive down to Phillip Island. On the way, we follow Stevo in his orange ute into a huge freeway-bound servo to fill up, grab some ice for the eskies and, as I suspected, hear him spray its concerned-looking forecourt with both aural barrels from his twisted farmyard menagerie via loudspeaker, making the whole servo think the console has been taken over by violent mutant pigs.

Like a kid with a fifth-grade science project, Stevo's spent weeks crafting a fake bottom in his ute tray's toolbox to smuggle in a whole Dan Murphy's of illicit booze, even lashing the excess bottles to his axles, and he dances a jubilant jig in his matching orange fisherman's pants when he gets it all in undetected.

It's three days before New Year's Eve 2006 and we've just made it into Pyramid Rock Festival. The summer sun is pouring fierce gold down from a perfect sky as we look for a campsite and four days of jubilant debauchery stretch long before us.

But a breeze soon picks up that quickly morphs into an absolute gale that rips Diablo's tent to pieces as he's trying to erect it. He gazes longingly at my one-man tent, and then at Stevo's three-man and, just like me at The Prom, I see the horror spread slowly across his face.

He somehow survives Stevo's pneumatic snores for two nights, but on the third night, when Stevo brings home a companion, Diablo is launched off the air mattress onto the rocky ground, a captive, harrowed audience for a new collection of eager farmyard noises. Not happy, Diablo heads into town the next morning to buy another tent. But all that's left are tiny kid's beach shelters with no door, which he buys and sits in, with gritted teeth, smoking joint after joint in a wretched, cursing state, while Steve's whooping laughs and evil cackles intersperse with long bows and tearful cries of, 'The Duke of Gumbletania, inside his *grand* manor.'

It's finally New Year's Eve, and Stevo has managed to force glowsticks through his stretched earring holes – two searingly bright orange connectors to match his pants, that join to form large hoops – and clad in a massive Amsterdam poncho-cape, he looks like a gypsy pirate superhero in the shifting light. His new Converse shoes were too tight so he'd sliced off the ends with a pocketknife. He calls them his 'Shongs', cackling gypsy-like as he wriggles his now free toes in the sun.

As I watch him dodge and weave among the hordes of freaks ahead of us, his tattooed hairy bulk twisting and cavorting through the fragrant air, he raises his 3-litre bottle of vodka and breakfast juice like a fat sloshing sceptre – a greeting to all who cross his path and a symbol of his freak royalty.

Throngs of shoe gazers, punks, rockers, wanderers, stoners, pillheads, beautiful freaks, layabouts, dreamers and maybe even a few Gumbletons march up and down the dirt roads between the stages, stalls and campgrounds in all their fabulous finery. Green witches, giant rabbits, purple-cloaked wizards, thin men in morph suits in every primary colour, German gas masks, inflatable pool toys, clown suits, a hairy-backed man in a red velvet conical bra – anything that fits in the weirdest procession imaginable.

I amble about the campsites making instant new friends, like I did when I was five, laughing at the random chaos that litters itself everywhere in tiny spot fires of hilarity and raging infernos of gleeful uproar.

Friendly new strangers invite me back to their campsite: a mega village of conjoined tents sheltered by huge tarps, awnings and portable gazebos, where eight dudes all in *Scream* masks are playing gridiron. I sit on an inflatable orange couch and am continually fed Sour Cream & Onion Pringles, Allen's red frogs and Stolichnaya vodka. I sit and smile, feeling the past two months of pain and sadness begin to melt away as someone rolls two perfect joints and passes them around our happy circle, smiling like a master craftsmen at the fruits of his labour. The smoke tastes sweet as the sun pours down from above, and the THC sparks up my body's cannabinoid receptors in a satin embrace, stretching my lips into a smile I feel inside every inch of me.

Soon enough, my new tribe gathers up provisions – two beach balls, one of the inflatable couches and two 3-litre casks of goon – and sets off in a shambling convoy toward the main stage, a simple six-minute walk that turns into a meandering half-hour hike as we stop every twenty seconds to engage another addled smiling stranger in rapturous conversation about nothing in particular. One of these strangers is a petite redhead with an English accent and huge, perpetually surprised blue eyes. She's lost her friends, so she tags along with us, taking a squirt of golden goon as her initiation.

She falls in besides me, and looks my way, studying me closely.

'You look rather lost in thought, my festival friend.'

'I'm just soaking in all the madness, keeping it in my head. To store it up.'

'I see . . .' she says, her brow furrowed in concentration. 'Well, I suppose madness *is* a rare commodity, after all.'

'Well. . . not *all* madness is,' I say, thinking I could tell a few tales in that regard. 'Just the joyful, sun-drenched, high-as-fuck kinda madness. That's the good stuff.'

'Agreed. But why do you have to hoard it?'

'Because tomorrow we go back to our crappy lives and I wanna remember the good stuff like this with visceral clarity, if I can.'

She stops walking and appraises me fully.

'*Visceral clarity*, you say?'

'Well . . . yeah. I suppose I wanna to write about it as well.'

'Ahh. And what's your name, *visceral* writer?'

'Well, I'm not really a writer. At least, not yet. My name's Dave. And yours?'

'I'm Julia. Lovely to meet you, Dave. Though I think I should tell you that sitting back and mentally recording everything is no way to be at all.' She takes my hand and dances me off into the throng in a looping waltz. 'You must join *in* the madness,

Dave, chase it, breathe it, *become* it! I assure you that *these* moments will be the memories you cherish.'

We dance, spinning like Julie Andrews in *The Sound of Music*, straight into a cardboard box robot, fashioned completely out of VB slab boxes, and by the time we apologise, take a few hits on his mate's joint, and realign his wonky box head, we've lost the gang. We're higher than nine hippies hotboxing in a helicopter, so we go and get some food. We take our quesadillas and coconut chai toward the water, where the sun's slowly sinking atop the hills to our west, lighting up the wilds of Bass Strait, rocks jutting out from her crashing seas. We sit there for an hour, as the swells far below shine silver and gold, throwing off endless bursts of light. We're talking about meditation. And Eckhart Tolle.

'So you discovered that nothing matters but now?' she asks with narrowed eyes, taking a final long drag on her thinly rolled joint.

'Yep. The more I meditated, the more I realised it. All the pasts and all the possible futures are just ghosts that haunt us in the present, stopping us from truly living. The more we spend wrapped up in them – and reacting to them – the more our ego drives us into what we *think* we want, but it never satisfies us.'

'Very eloquent, and very *Zen*, Mr Writer, but didn't I hear you less than an hour ago moaning about having to go back to your crappy life tomorrow?'

'. . . I'll pay that.'

She raises her chin in mock victory. 'I should think so. Plus, we all have to want *something*. Otherwise we'd be pointless slugs without any desire or motivation.'

'Pay that too.'

'What do *you* want, Dave?' She tilts her head in self-appraisal, then cracks up laughing. 'Oh my God, that's so pathetically clichéd! A real Hallmark moment. "What do **you** want, Dave?"' she mocks herself in a high-pitched sing-song voice.

'Well, firstly, you *are* very high, so a certain amount of slack must be cut.'

She considers this and reluctantly nods.

'And secondly, it's not only a perfectly reasonable question, but a damn powerful one.'

'Hmm?'

'Well, what do *you* want, Julia?'

She grins and shakes her head.

'You first, Dave. *I* ask the mortifyingly corny question in the orange rays of a clichéd ocean sunset and *you* give the corny answer. No messing with the script.'

'Right. In keeping with the corny theme, I want freedom, Julia. Sun on my skin. Good magic, amazing music. I wanna hurl myself off to far-off corners of the globe, to meet beautiful women on foreign beaches.'

Her eyes narrow sardonically. 'So, essentially, you want to be James Bond?'

'Very funny. No, I just. . .' I stop for a second and, through the blissful haze of the weed, actually consider what it is I do want. 'To be honest, my life kinda sucks right now, so coming out here is the first step in reclaiming some joy and squeezing some magic from my shitty life.'

Her lips turn up in a small smile. 'And how's it working out for you, so far?'

'What, magic?'

'Yes . . . And the rest.'

'Well, the magic, freedom, sun, joy and music are all big ticks. No foreign beach, but this incredible view more than suffices.'

'And the beautiful women?'

'I'd say one is more than enough for now.'

'Ahh, very Bond.'

Her eyes are huge, surprised and gorgeous as I kiss her, and my God it *is* clichéd, but it's also incredible, and sometimes – very fucking rarely – life fits snugly into that Venn diagram, and when you're high on amazing sativa, atop a cliff on New Year's Eve, basking in the hazy-orange gold of an ocean sunset, kissing a beautiful girl with not a grifter or hobo-freak in sight, that's a Hallmark moment that's more than okay.

We head back through the crowds that thicken, metre by metre, toward the massive main stage, light and sound pouring out like sacrament as 15,000 worshippers leap and twist in a teeming, chaotic mass of joy. Every member of the inflatable animal kingdom is leaping about, propelled by a million hands, bouncing through the air with a thousand beach balls.

We're deep in a leaping forest of smiles and kindled eyes. The weed enchants everything, coats it with wonder, and the stage pulls me in like an astral plughole. Fierce warm gold presses out from the stage lights, washing over endless faces drinking it in. It washes through my pupils and sloshes against the back of my skull like a glinting wave against the rocks. The blue of the sky is deepening as fingers dance over strings and guitars are wailing, looping, crying like swooping gulls. We are highly invested in immediacy.

We dance for hours and before we know it, it's a minute left until 2007. Everyone leaps about like excited puppies and two, one, zero, a roar grows from the bellies of this army of now and I *bask* in it, smelling the salt, the dirt, the nachos, the sunscreen, the bourbon, the weed and the joy I've been hungering for that permeates everything.

Later on, I find Stevo, propped close-eyed against the back wall of the music tent, dirty electro beats juddering from its DJ booth as scattered ravers bobble and sway in the twisting light. He's higher than Jesus, and drunker than Satan, his 3-litre vodka-breakfast juice empty and a still unfinished joint in one hand, a soft serve cone in the other. He raises it to his mouth and *misses*, adding to the smeared ice cream already there, then slowly drifts off into stoned sleep. I crouch down and pilfer the joint, and as he drops the cone, I place a gentle hand on his sunburnt shoulder.

'Dude, it pains me to say this, but you might just be a Gumbleton.'

One eye jerks open, then he snorts and slowly drifts off again.

Julia leads me back through the wilderness of stumbling drunks and munted ravers clad in garlands of fading glowsticks to her five-man tent. It's scattered with cushions and has a mandala tapestry somehow hanging from a wall. She lights a stick of frankincense, then pulls out some dark chocolate and a half-empty bottle of wine, pouring some into a cup for me. She bites down on a square of chocolate, blue eyes on me, and I sip my Merlot as the frankincense creeps into my nostrils.

'You seem deep in thought once again, my festival friend.'

'I'm just . . . *impressed.*'

'So you approve of my humble abode?'

'I don't think I've ever been inside a classier establishment.'

'Ahhh, flattery may just prove advantageous to your cause.'

She crawls toward me slowly, takes my cup of wine from my hands, sets it down and pushes me onto her inflatable mattress. The distant music is twisting and thumping from mainstage, and we kiss, long and slow.

My hands swim down her sides, and my lips to her neck, brushing the gossamer-thin hairs there, and she kisses me hard, biting my bottom lip as her pelvis pushes up against me. The music peaks as I unroll each stocking, her eyes hungry and conniving, her red hair splayed out against the mattress. She shrugs out of her underwear, her hands greedily guiding me in, and she murmurs, twists, wraps her legs around me and pulls me deeper, and I'm falling now, in and out of wonder and perfect lust and all the delights that come with freedom. In the rising strings and thumping bass, she cries out and I explode, in an endless second that elongates, rings like a bell, before fading, as I sink back to 3D and lie there utterly spent, at peace and joy with the world and everything in it.

'Well, *that* was a Hallmark moment!' she says.

We laugh until our stomachs hurt, and trade small kisses as we lie in this nylon cocoon, staring up at the shadows dancing on the roof, listening to drunken curses of those outside, tripping on tent ropes.

She sits up and stretches like a cat, arching her back as her fingers graze the top of the tent.

'You know, now I think about it, some of the best sex I've ever had has been inside tents.'

'Really? I can't say I'm all that experienced.'

'*God*, you probably think I'm some whorish strumpet now!'

'I think you're a fascinating, accomplished and fucking gorgeous woman.'

'Ahhh, more flattery means more gifts.'

She climbs off the mattress and hunts through a leather shoulder bag, producing a small lump of pressed hash. She works it slowly in her fingers, rolling it into a long thin noodle, then takes out some weed, a grinder and some papers, and rolls a picture-perfect joint around it. I stare at her nimble fingers, even more impressed than before.

'You're a keeper.'

She feigns a look of shock. 'High praise. But we're dealing solely in *now*, remember, Dave? *Now* is where all the fun is.' She drinks more wine and so do I. We finish the bottle, then light up the joint and the rich sweet smoke mingles with the frankincense.

The hash unfolds its wings. The taste of the chocolate is sex in itself. We lie back on the cushions, and I float up, tracing hidden sunlit skies. She floats too, and after a while I find her, and we twist together like helixes of light. We face each other open-eyed, tracing fingers down each other's sides. Hers feel like shooting stars.

Smoke hangs in the air in drifting swathes as I fall gently into a stoned sleep, dancing free and joyous somewhere, as orange suns kiss golden seas and I soar up through lemon cloud to look the heavens in their silver faces. And on until morning.

The sun peeks above the waves and begins to cook the tent, so we wake, drink some coconut water, and say our goodbyes. I drift with a head of velvet dream back to our campsite and get ready to pack things up for the long drive home, smiling in both the infancy of the day and its new year, now stretching out in potential wonder.

I made the most of those weekends, because as much as I tried to take that magic back with me, as soon as they were over, I was back to being Hades' ringleader, micromanaging all my suburb's petty fuckery and hornswaggle while still clinging to the last vestiges of my sleep, patience and sanity.

A few weeks into 2007 and things were going swimmingly. Our staff toilet was refusing to flush again, just after I'd made an explosive, sizeable sacrifice to the porcelain gods, which I was doing much more often, with my burgeoning gut problems. I knew my servo diet wasn't helping matters, but I'd be serving customers and would suddenly double over in pain, hunched in helpless agony until it eventually passed. Doctors were no use. Naturopaths weren't much better.

Our aircon had died, and it was one of those sweltering, full-moon nights where everything was lit up like Vegas and even buttoned-down accountants who still played World of Warcraft ripped off their shirts and howled in the strange silver rays.

The store was a rollicking Louisiana houseboat on PCP, rolling up the dark Mississippi. Sirens pierced the night but still a conveyer belt of Commodores, Skylines and Falcons sold ounces of drugs in the carpark as Randall approached them hopefully, strumming his guitar and trying to bum smokes off them, mid-deal. The slurpee machines had died and a strange silent goth girl was periodically meditating in the ice freezer while spiky peroxide-haired arsehats with red-rimmed eyes leapt out of long-suffering taxis yelling rugby chants and scoffing endless unpaid chocolate bars, letting the wrappers drift to the floor like autumn leaves. They slammed into the glass doors and laughed at the reverberating rattle.

One of them decided he wanted a 'hot' Coke and put the can in the microwave on high for five minutes, before running off into the night, chuckling to himself. His mate threw his guts up in front of the gas bottles, then crouched down to get a picture of it, circling it repeatedly to find the best artistic angle. Randall approached him furtively, asking to trade a song for a smoke and as he played some basic chords, the guy sang 'Wonderwall' over the top of it, then gave him half a packet of PJ Gold 30s for his troubles. Randall's face lit up like a kid at the circus, and there was me inside, wondering if I really could manage to live permanently on Centrelink after all.

Before the spew was dry, in drove Marty – straight over it – in his black XR8, psytrance twisting out the open windows in aural figure-eights, a doe-eyed blonde in the front seat, sucking on a baby's pacifier. He burst into the store, bouncing and shuffling to the alien squelchings and machine-gunning beat, grabbing eight packs of Extra, eyes black dinner plates, as usual.

'Daaaaave, my *brother*! What's kicking?'

'That beat, for starters.'

'Hahaha, fuck yeah! Came in for a chewie run.' He pointed to the front seat. 'It's her first time and these things are so strong they're making her chew her lips off!'

Turning himself to block the view from the cameras, he surreptitiously took a small baggie out to give me a squiz.

'They're called "Blue Skies", 'cos that's where they take you, brother. Straight up into the big deep blue.'

As it turned out Marty had an ulterior motive for dropping into my store, one which actually managed to put a smile on my face. 'I'm renting a beach house in a few weeks,' he said. 'Reckon you can get some time off?'

'I'm in.'

Back to Neverland. I'm still not sure who'd entrust their luxury real estate to Marty, even for just six days, but it was a stunning place, in Torquay, right on the edge of a manicured golf course, and with several close mates, quality tunes, an ounce of Northern Lights, a wild surf beach and a huge balcony overlooking the 18th green, we were set.

Sitting there watching the tanned baby-boomers teeing off like some recreational conveyer belt, I slowly drained Stella after Stella as the sun set over my head, lighting up the dunes and meandering greens. I felt that familiar peace wrap itself around me again, as Moby's 'Porcelain' leaked out through the open door and onto the balcony, where me, Marty, Stevo, Diablo and others pondered the many meanings of life, of peace, of Gumbletons and of Cheezels, as we casually popped them off the ends of our fingers and into our mouths, washed down by sparkling Belgian beer.

As the sun fell, we stole out onto the famously designed course, mobs of roos bounding away from us up the bending hills and valleys, the orange and purple sky soon pastelling to roving cloud silvered by a fat full moon. For two hours we hunted Titleist golf balls – like wide-eyed kids on Easter morning – spotting their pearly luminescence dotting the drunken hills under that midnight sun, the green-black world teeming with croaking frogs and crickets.

The sheer joy of the whole thing sticks with me today, running up the 7th fairway, as it curled up toward the top of the dunes, pockets bulging with balls as I stopped, breathless and baked beyond all belief, to take in the view of the silver waves crashing ghostlike onto the shoreline below. Something overtook me right then, something numinous slipping in again. I sat there for an hour and a half, my body glowing with pure joy that filled me so much it spilled over, tears leaking down my face.

I never tired of watching those waves, their distant roar in the supernal gloom as crests of foam caught the moonlight on their endless cycle of surge, break, crash and glide, followed by another, and another, and another, for all eternity. Mother Earth's great hulking heart was pushing them out onto sands all across the world in endless patient love, entreated by that luminous magnet in the sky. I followed those crests out with my eyes, and a yearning filled me so powerfully my breath caught in my throat. I *would* travel over those waves again. I'd get out of that fucking cage and into the guts of a big metal bird and find my Neverland. As sure as that glowing rock in the sky encircled the earth. And this time I might not come back.

When they eventually found me, shining the torch on my tear-stained, dumbfounded face, the spell finally broke and we all headed back to the house for more weed, movies, Tim Tams and Cheezels, our bulging pockets swinging in the breeze.

These adventures made my servitude bearable. Me and all the nameless soldiers toiling in their retail deserts of aisled halogen all over the country and all over this gargantuan rock jetting through the yawning blackness of a glittering void stretching far beyond even the faintest reaches of wild imagination.

We need our breaks, for we Servo Daysleepers are a delicate but advanced hybrid. We're picker-and-packers, check-out chicks, counsellors, sales assistants, MCs, philosophers, security guards, debt collectors and reluctant ringleaders.

Like cacti, we're resilient. During the day we lie dormant, soaking up the heat while the sun beats down mercilessly. But when the sun falls, we awaken, sprouting a multitude of brilliant pearl-white flowers that, fragile in their infancy, tremble toward the lonely moon. At dawn these flowers will be dead, killed by the power of the sun's searing rays. But they will be replaced the following night by more; by those who can only find comfort in the serene velvet of the night.

CHAPTER 19

SERVO KARMA

I've never been able to charge customers for sauce. I just can't do it. It's stupid, I know, but to me it's emblematic of how cravenly mercantile this world has become, where a national airline removes a single olive from their in-flight meal, saves $40,000, and now even a fucking toothpick has a price tag on it.

It all started early one winter morning back in my first year as a new dawn hovered at the cusp of the world. A cold wind – redolent of diesel fumes and wafting fat from a nearby fast food joint – blustered through the dirty forecourt giving life to empty chip packets and plastic bags. A cold, haggard-looking man stumbled in through the doors, buffeted by the July wind. Not bothering to exchange pleasantries, he shivered his way toward the pie warmer, where his puffy eyes lit up at the wealth of golden pastry.

Basking in its reflected heat, he rubbed his hands and grinned like a kid looking under the tree on Christmas morning as he lingered over his choice. Grabbing a steak, cheese and bacon pie, he approached the counter, jingling several coins in his pocket. Upon reaching me he chucked them onto the promo mat on the counter, greeting me with a warm, satisfied smile.

'G'day bloke. Fuck awful night we're havin', ain't it?'

'Sure is.'

'I've been dreaming about this brekkie for four and a half hours.'

'I hope it's all you imagined.'

'Yeah . . . *it will be*! Don't you worry about that!'

Although he'd already paid for the pie with his small pile of coins and was clearly starving, he hesitated.

'Ya got any sauce mate?'

'Yeah, sure. That's twenty-five cents.'

Even before I saw the colour drain from his face, I knew I'd made a monumental mistake.

'Ya takin' the piss, mate, that's ridiculous. Ya can't charge for sauce! That's . . . *unAustralian*!'

'But that's what it costs,' I said, already weakening.

'Look, mate, this is kinda sad, but I ain't got an extra twenty-five cents,' he said, as that same tired look stole over his features again. 'I've robbed all me couch cushions and me centre fucken console, and that's all the dosh I got on me 'til I get paid tomorrow. I'm as broke as a joke.'

'. . . Oh . . .' I said, downcast, before a realisation suddenly hit me. 'Well, I tell ya what, I'll just give you one. Actually, stuff it, have two. And . . . you know what? I'm about to put some new pies in soon, so you can take the other two for free if you want. And here's some sauce for them too. It's not like they bloody count 'em anyway.'

His face lit up like a beacon.

'You're a fuckin' champion, mate!' he cried as he seized my hand and shook it furiously. He grabbed the sauce and the pies then headed out the door with dangerously moist eyes. 'You have yourself a top bloody night!'

And so began my odyssey to supply every worthy customer with complimentary sauce. And pastry, and slurpees, if they really deserved it. Over the years I extended this into a kind of ghetto religion: Servo Karma.

My forays into meditation had shown me not only the peace that could be attained, but they also sharpened my senses. I soon began to really *see* my customers. So many of them shuffled aimlessly around, flung here or there by unseen strings, they seemed to be eking out Dickensian lives of quiet desperation. So often I stood with one eye on my forecourt and the other on their trundling frames, wondering what could truly emerge from the wellspring of their swirling psyches, if only gifted a sincere pinch of passing kindness.

The more I observed them, the more I found my servo to be the perfect environment for the establishment of my new creed. Servo Karma is, I suppose, my small protest against Orwellian avarice, loosely based on certain principles of Buddhism, enmeshed with smatterings of decency, me being in a good mood, and common-fucking-sense.

As a retail religion, it has many facets, but what I was essentially doing was extending my newfound spirituality and coalescing it with mom-n-pop store techniques in a much larger, staler environment; where such simple empathy, generosity and common sense shone like a beacon into the murk of bureaucracy, mercantilism and rapacious, institutionalised greed. People loved it because it showed them that decent human beings still existed. I loved it because it meant I wasn't one of the bean-counting arseholes. The following is some of the ways I did it.

Ask any young retail worker and they'll tell you that at almost every place they've worked at, the till would always have to balance. If it was down 5 cents you'd have to explain why. While servos obviously care about where their money goes, they're not as strict on variances as your average supermarket. The breakneck speed of the place always leads to cash-handling mistakes, especially with new staff, and it's a rare event when somebody has a variance figure that's $0.00.

Not once did I worry about balancing my till. I mean, I wasn't like the crazy guy in Munich who charged what he wanted; I processed all the transactions correctly and accepted legal tender. I just . . . let things slide sometimes. I'd let people down on their luck off 50 cents, even five bucks if they really needed it. The thing is, I'd always get it back before the shift was over. Loved-up clubbers on pills would come in at 2 am and buy packets of chewie, giving me a $10 note and saying, 'Keep the change, bro.' That's Servo Karma in action, specifically John Lennon's 'instant' variety.

But it was more than just money. I'd call taxis for people. I'd help find them a hotel. I'd play Dr Phil for troubled relationships and offer advice and hugs while listening to sob stories from the broken-hearted. I wouldn't sigh exasperatedly at people thumbing through magazines. I'd smile at people and actually *mean* it. I know these seem like simple everyday things – and they most certainly are – but so many servos you go into these days are manned by sullen students, worried robots or bitter people looking for a fight. I think my main reason for spreading

Servo Karma was something Rathindra, along with Confucius himself, once said: 'Be kind, for everyone you meet is fighting a hard battle.' Particularly those with body clocks broken enough to be animate at 3 am.

Servo Karma would always pay me back in droves, and always when I least expected it. Twice, I was driving to work, late, angry and gunning the accelerator through empty roundabouts when my heart sank as the inside of my car was bathed in twisting red and blue. And on each occasion, I was met by a stern-faced officer who, after asking why I was speeding, would notice my work uniform, then my face, then remember my brilliant service and say, 'Oh, you work at the servo, don't ya? Off ya go then. Bit slower next time, eh?'

Even better, though, was one night when I was checking used-by dates on the milk when a crazed young man in double denim and a long rat's tail powered into the store, butchering 'Satisfaction' by The Stones at the top of his failing lungs. Off his face on booze and coke, he walked straight to the ATM as I slipped behind the counter. After he got his money, he fixed his black eyes on mine and strutted toward me. He then slammed a $50 note down on the counter, and stared at me for five whole seconds, reeking powerfully of aftershave, and said, 'That's for you, mate.'

He explained that over three years ago, he came in begging for a condom and I gave him one.

'I was more toey than a Roman fucken sandal that night,' he said, his sweating face deep in the ghosts of Christmas past. 'I'd been at this party, and this chick, she's up for it, right? Absolute stunna. But how's this? I didn't have a fucken dinga on me, so I sprinted to that other servo down the road and asked ol' mate to spot me a pack, just 'til morning, but he wouldn't gimme the steam off his own piss, so, desperate, I came in 'ere, and youse had no fucken dingas in stock! I thought it just wasn't my night, but you gave me one outta your own fucken wallet. King of kings you were that night, mate. An' that was me first fucken root! So buy yaself somethin' nice!'

And I did. Although at the time it was probably acid or pills. Anyway, as time went on, I began to find many more spheres to spread my new religion's growing influence.

Some were ridiculously simple: One thing about servos I'm sure a lot of people are sick of is that every time you approach the register, you're asked if you want two king-sized Mars Bars for $5. It's annoying, I know. If you really wanted the Mars Bars, chances are you would have grabbed them yourself. I want every reader to know that in keeping with my Servo Karma doctrines, bar one aforementioned experimental night, I've *never* pestered anyone with that insidious crap in all my time serving.

Because of my reluctance to harass people, I didn't do too well in my sales figures. You see, every time a console operator bombards you with that insufferable crap, they've got dollar signs in their eyes. They're foisting chewie, chocolate or Tic Tacs onto you not only 'cos they're a pathetic simp addicted to the head-patting graphs from head office displayed on the employee noticeboard, but because they see themselves as in with a fighting chance of some tiny gift-card crumbs that get occasionally brushed from our benevolent masters' table.

The percentage figure of customers that gave in to our pestering and bought whatever assorted crap we were trying to get rid of at the time had to be 8 per cent at a bare minimum. If the figure was less, you were 'gently encouraged' to do better and if you didn't you were soon in trouble. While our region encompassed ten stores and over 200 workers, I was often dead last for my badgering results. It's probably my proudest achievement over the six years and arguably the most integral pillar of Servo Karma. Hilariously enough, as time went on, through some unintended reverse psychology of not asking anyone ever, I ended up selling more and more promo deals until I almost topped the charts, spending my gift cards while laughing victoriously to myself at the Universe's unique and reciprocal sense of humour.

Another is that I would almost always tell customers the truth. While this may seem like a recipe for disaster, I'd often find that it would usually turn out in a distinctly different way. Getting asked hundreds of times on every shift, 'How are you?' I found being honest would actually refresh some people. Saying I felt like shit spread on burnt toast at least gave people an unexpected belly laugh and engendered some connection and humanity, something I felt was rarer than rocking-horse shit under that staid and mercantile desert of upselling and staged lines.

One night, when I was seriously struggling, a customer asked me how I was, only for me to reply offhandedly, 'Fucked.' I looked up and into the eyes of a sprightly old woman whose eyes lit up as she said, with a naughty grin on her face, 'Me too, dear! Me too!'

As the nights continued on, with me needing some creative respite from uni's onrushing tide of focus groups, branding campaigns and endless marketing plans, I began to think of what possibilities I had of indoctrinating others into my newly formed religion. As the workers surrounding me continued to come and go, I began analysing them and seeing if there were any that were worthy of learning then spreading the glorious creed of Servo Karma.

I tried firstly with Andrzej. He was a tough case, because he already had his own system of treating the customers worse than they treated him, only with an extra dollop of malice on top, hopefully intimidating them into taking their business elsewhere, leaving him more time to drink scotch and read the works of Hemingway and Harper Lee.

The problem with such a system is that karma – servo or otherwise – is an unremitting two-way street of universal reciprocity. Andrzej forgot that, when pushed, a small but meaningful percentage of society can be vengeful fuckers who enjoy seeing the bug they are tormenting thrash around with its little legs in the air. Powered on by Andrzej's challenges, they'd return, night after warring night, trying everything they could – from continually filling their cars with 45 cents worth of petrol every hour to simply walking in and pelting eggs at him – to finally crack him. I tried in vain to show him this vicious cycle, and urged him to try meditation, but I never had any luck.

Undaunted, I started on Nellie, but she was even more set in her ways. I was standing outside with her one quiet autumn midnight in 2007 as she smoked her Winnie Reds and I munched on a sausage roll, trying to espouse the most appealing facts of Servo Karma as I squeezed more pilfered sauce onto every bite. After listening to my spiel, she took a long drag on her cigarette and, looking me squarely in the eyes as a fuel delivery truck groaned in, said: 'Fuck them all, David. To the moon and back.'

But I remained convinced I had a reasonable heuristic for life behind the console. I told El Diablo about Servo Karma and he thought it a noble and glorious thing.

'Keep it up, man, we need more resistance in this world, helping brothers and sisters out against the fascist fuckers. I'll join your revolution, bro!'

But I needed more soldiers where they could make a difference: in the trenches. As I started to get desperate, I thought to myself, why not turn to the service station industry's greatest natural resource – Indians!

I began to befriend and then teach this new guy named Rakesh. Rakesh's main lesson from me was not to be a company-approved automaton and to engage people in casual, humorous and somewhat irreverent conversation so they didn't treat him like a drone. So many Indians I knew over the years ended up the butt of every second customer's jokes, with a not insignificant dollop of this being straight-up racism, coated in a thin veneer of jocularity. With Servo Karma, I resolved to change this and stop the abuse and its ensuing running battles that were waged every night.

So, just like young John Connor did with Arnie's cyborg T-800 in *Terminator 2*, I began to teach Rakesh to relax; to artfully utter Aussie profanities with expert timing; and to stand up for himself, but tactfully, and with humour – so as to not end up a harried mouthpiece of the company who got into a fight with every second tradie who he wouldn't let off 10 cents for their diesel. Soon enough, my student began to progress. I flushed with pride as he began slouching, telling the unvarnished truth, assuring customers that 'she'll be right, mate', giving out free pies on the sly, and even voicing his doubts about stupid company policies, as he peered to the left of his large frames to see if I approved.

But now and again he and others would falter as the callous rudeness and disgusting racism of the customers got to them. I discovered that, eventually, they *did* crack. When some witty bogan strode in and bellowed at the top of his failing lungs, 'What happens if I scratch that dot off your head, mate, *DO I WIN A CAR?*', their eyes narrowed as an invisible switch flipped inside them, turning them from Mr Prim Proper and Respectable into Mr Let's Go, You Fucker.

They'd stare the Gumbleton down for a full five seconds, the whites of their eyes like raging phosphorus, then they'd seize the metal pipe from under the counter, weighing the heft of it in their hands, their voices jumping two octaves as they reacquainted themselves with their warrior state. Only when the conflict had passed did they gulp a deep breath, return the pipe and tuck their shirt back in, before putting their mask back on and smiling, with teeth, at the next approaching customer.

CHAPTER 20

EVOLUTION

While Servo Karma was making my work nights more fulfilling, away from the console I wanted something more transformative. Meditation was great and all, but its insights had slowed to the pace of an elderly turtle with the handbrake on. I craved a fast, shining shortcut; a more powerful inner exploration to strip back the many layers of egoic false identity I'd accrued over the years of button-pressing and freak-herding.

While most would consider this a hefty and probably indefinable task, to El Diablo, it was an answer enclosed in seven solitary letters.

I saw his lumbering green station wagon, named Betsy for some inscrutable Diablo-flavoured reason, pull up in the rain one cold winter night, and as the auto-doors slid open, he raised a finger to his lips and pointed at the sky.

'See that, Dave?'

'Not really, it's just blackness.'

'No, bro, *the rain*!'

'Well, I can hear it, at least.'

'It's like gold, brother! Gold!'

'What am I missing here?'

'*Shrooms!*'

'Shrooms?'

'Yeah man, they're your final frontier! And I will be your guide. We'll go picking soon when we get a bit more of this beautiful gift from the heavens.'

And out he went to dance in it, backlit by the canopy lights as the drops pelted into him, arms raised to the clouds in thanks and benediction. A buttoned-down guy on pump five with a haircut you could set your watch to shot perturbed looks at Diablo as he hunted through his discount dockets, then gave him an extra-wide berth on his way in to pay.

Two weeks later, after several solid days of rain, we set off for the Dandenong Ranges early on a frosty morning – me with a glorious weekend off, and Diablo, Marty the Party, and various camping supplies all packed into a sickly, spluttering Betsy – Diablo coaxing her on by whispering sweet nothings into her air vents – as we headed for the fungus-ridden hills.

On our way there, Diablo gave me a crash course in shroomology.

'Dave,' he grinned as he weaved through merging traffic, an addled SpongeBob hanging from the rear-vision mirror, smoking a bong but seemingly enraptured in our conversation. 'What can you tell me about shrooms?'

'I, uhh, know they're psychedelic, that they work on the serotonin system and—'

'None of that neurochemical nonsense, my brother! *Shrooms are a radio!*' He turned up Triple J to elucidate his point.

'A radio?'

'That's right! And what does a radio pick up?'

'Uhhh, radio stations?'

'And what are *they*?' he asked patiently, whipping his head side-to-side to the indie-electro, the car mirroring his movements as horns blared and SpongeBob swung like a giddy metronome.

'. . . frequencies?'

'Frequencies brother, exactly!'

'Whose frequencies?'

'The Mushroom Gods, man! That's who'll be helping us in the hunt today. If they want us to find them, then we will. Today we enter into a holy communion, not just with Mother Earth, but also the sacred radio station of the Gods.' He blasted the volume and bopped his head furiously.

'But what if we don't find any?'

Marty tapped me on the shoulder and grinned, showing me a goody-bag of MDMA and acid as a rather solid back-up plan.

Just like searching for midnight golf balls, mushroom hunting is a grown-up Easter Egg Hunt. You feel like a kid, excitedly finding those tiny solid chocolate eggs all over the house, shining like precious emeralds and rubies in their foil wrapping in the soap holder in the bathroom, in the teacups of Mum's best china set, and, somewhat fittingly, inside the egg holders in the fridge. The bonus is that instead of a sugar-theobromine rush, you now get psilocybin, a skeleton key to open the door to psychedelic alien wonder.

After a quick mushroom spotting lesson from El Diablo – 'the ones we want look like this, but the ones that will kill you by turning your liver to mush also look like this' – we started as the wan winter sun climbed over the tops of the canopy. I squelched through the damp earth as we searched, my feet falling through rotten logs and past legions of ants and other bugs feeding on the rich decay of Nature's cycle.

As I followed him, Diablo turned and shooed me away in an opposite direction so I didn't waste time covering the same ground. I headed the other way, down into a valley, the laboured breathing and tiny woops of El Diablo and Marty dying away as I explored further, eyes darting to and fro.

I searched for the better part of an hour, reaching the bottom of the valley and then stopped dead still, almost tripping over a blackened log. Before me, among a cross-section of fallen eucalypt branches, lay hundreds of gold-topped mushrooms. I wasn't sure if they were psychedelic, poisonous or both, but they were enough to either switch on or wipe out an entire army.

I let out a yawp, which quickly drew Diablo, crashing hopeful through the undergrowth, his manic grin almost touching his ears as he laid saucer-eyes on its magnificence, plucking one up and soon showing me the blue bruising from the shroom's ample psilocybin waiting inside, loosing a rabid 'MEOW! MEOW!' like some starving feral cat happening upon a bounty of tuna.

Diablo all of a sudden went solemn and still, putting his arm around my shoulders and bowing his head to the wet earth.

'We give thanks to you, Mushroom Gods, for this glorious bounty you've bestowed upon us. From whatever galaxy you awesome fuckers came from, we're grateful you chose this planet to help raise our consciousness. Me and Dave are honoured by this gift.'

I was touched, but also cognisant of the fact that too many shrooms could hand the universe a whisk to fluff my callow cortex to a pink-grey mousse, and thus I resolved not to overdo it.

We met Marty on the way back to camp, whooping triumphant, our arms cradling countless paper bags full to the brim. Diablo emptied out a bag, eyeballed them then divided them into three equally hefty piles he placed on different coloured plastic plates. I glanced at them nervously.

'Umm, shouldn't mine be a bit smaller than yours? First times and all . . .'

'I usually have three times that amount.'

'Oh.'

'You'll be fine, my brother. As long as you surrender to them, the Mushroom Gods are gracious and wise.'

We gobbled them up and waited for lift-off, listening to music and talking. The sun had broken through, coating everything in a lemon glow, and as I looked up to the tops of the gums, even though there was no wind, the trees were waving. No, not waving . . . *breathing*. Everything around me was breathing. It was sentient. *Alive*. There was actually conscious life twisting up through the bark and into every quivering leaf. And down through their roots, pushing energy and information through the unseen highways of the rich earth. I felt a *droopiness* spreading up from my guts, through my chest and into my head. The world began to dim and blur at the edges like a fish-eye lens, the sky pulsing between glowing cerulean pastel and a deep midnight blue. Sinking into the camp chair, I felt as though some unseen force was taking an eraser to the outline of my being, rubbing out the barrier between me and Nature. As it did, Nature poured in, and my mouth fell open as I realised how much I was a part of it all.

I stood up, testing my legs, and headed out of the clearing.

'Dave.'

I turned, and Diablo handed me a water bottle.

'Safe journeys, brother. Go well.'

I headed out, drawn by the line of gums ceding into the blue-grey of the sky. The huge rocks, the rotting logs, the legions of ants trekking over lush carpets of green moss. I followed them, like a hobbit who'd seen strange lights deep in the woods.

Soon everything started to billow and walking became impossible. I came to a giant gum and sat down with my back to it, immediately melting into this small corner of the ecosystem. I could feel the tree through my back. Vast, powerful, selfless and giving. Here it was, dragging in carbon dioxide and pouring out breathable air for decades on end, while I simply stood behind a console pressing buttons and smiling sardonically; it had a better résumé than me.

I took in the deepest breath, held it, and let it out in a long slow gush, realising that I was the tiniest, most inconsequential speck of organic matter in a vastly intricate system of mind-bending precision and wild beauty. And that I was completely okay with that.

And in that wave of upswelling gratitude, the shrooms pressed fierce, far exceeding anything I'd felt so far. The world as I knew it faded away, the fish-eye lens pressing in rapidly to a pinpoint of light. That light then began to grow, the dot stretching into faintly glowing lines, its sheen washing slowly outward to show giant trees sprouting up from seeds to maturity in mere seconds, ripe, full and powerful, dropping more seeds to repeat the process again. My cells were green spinning fractals. Everywhere was growth and death, writhing and dancing under a pouring sun. It beamed down codes of light into every living thing, driving the growth, the development, the evolution. Then it hid, as skies darkened and rain poured down in sheets. Mushrooms sprouted up from the soil, glowing faintly but with potent alchemy. Animals were birthed, growing from eggs and infancy to maturity, birthing, then wasting away into death. Fish. Snakes. Birds. Apes. All continuing the cycle of rebirth, again and again. The apes lost their hair, little by little, and then they were human babies, adults, then elderly, then, in between death and decay, a great swathe of light escaped them, before new babies emerged to repeat the cycle again. I felt my heart swell.

I surrendered to it, and my inner world shot up into the vast sky as I looked down on the green and blue ball of Mother Earth, hanging fat, beautiful and impossible in yawning wastes of nothing.

Energies shot in through my crown, like a three-minute lightning bolt, jack-knifing revelation through my synapses. They felt like rivers of codes; like skeins of amethyst-coloured light full of holy life; like downloads from Somewhere Else, all shooting down my spine, my arms, my legs, and then snaking down slower to my guts, where they started . . . rearranging things, fixing things, somehow. I was being cosmically healed by . . . something . . . a long way away.

I came to, lying on a bed of moss, branches and rich earth, the smell of it deep in my nostrils as I pushed myself to a sitting position. The sun had emerged more fully, striping the ground into a luminous tiger's pelt.

I felt in my pocket for something that was digging into my upper leg and found my iPod. I fished out its earbuds, switched it on, and into my head swam John Lennon, journeying across the universe. It sounded like slow showers of violins cascading down through the air in ribbons of luminescent colour. I got up, following the sun in the sky, basking in its warmth and light as John sung about universes, walruses and strawberry fields.

The gums were chandeliers now, as the sun deepened, pressing the shadows long and shooting honeyed spears through the canopy to cast a dappled spell over everything. I walked for an hour, stopping to sit a few times, as my breath fogged in the cold air and the sky darkened. I never panicked, knowing the forest would guide me home. And then, in the last of the gloaming I came upon the campsite, a fresh fire bouncing its hot gold into the air. El Diablo sat, hooded and staring into the flames, finishing off a joint then tossing it into the fire. He looked up.

'How long was I gone?' I asked him.

'A long time.'

'Forever?'

'Forever *is* a long time, buddy. It stretches out, like a highway of eternity as we rocket into the great laughing unknown.'

'Where did that come from?'

'Mushroom Gods.'

Marty emerged from the tent with a can of Sour Cream & Onion Pringles and a packet of Allen's Snakes, gifting them to me with a crazed but munificent smile. While I wished I had something healthier to gift my rearranged guts, they tasted transcendent. I thanked him and noticed his pupils were the size of tennis balls.

'You look like you had four times as many shrooms as me,' I noted.

'Hippyflippin, bro.'

'Hippyflippin?'

'MD and shrooms.'

'Ahh . . .' I laughed, almost choking on a red snake. 'That'd do it.'

I ate in silence. The dark was deepening now, and the roar of crickets, frogs and other denizens of the undergrowth was reaching its crescendo. Diablo watched me closely, the fire bouncing its light onto the surrounding gums and dancing orange light over his hooded face.

'Well, Dave, did you meet them?'

'I found the radio station, yeah. They . . . helped me, I think. They fixed something inside of my guts that been bothering me for a while. It's like I was being cosmically operated on by . . . light, if that makes any sense?'

His face lifted in recognition.

'They did that for me once, too.'

'They also sent all this information down into my head and through my body, like they were teaching it to me to make me . . .'

'Evolve.'

'Yeah. That's exactly what it felt like. How'd you know?'

''Cos mushrooms are like us, Dave. More than we know. They breathe in oxygen and breathe out carbon dioxide. And over time, they evolve. And they want us to as well.'

'But evolve into what?'

'Who's to say, buddy?' He arched his eyebrows. 'What about you, Dave, any evolution coming up for you?'

'Well, I still feel like a human at this point, but we'll see.'

'No, I mean in your life. You always tell me how shit it is at your work. How long are you going to suffer there for? Stuck in your prison? Evolution is change, brother. Remember that.'

I chucked a green snake at him which he tried to catch in his teeth.

'Thank you, José. I will.'

And I did. At least until Marty appeared from nowhere, like some deity from the clouds, beaming, offering me a fresh steaming bowl of mac 'n' cheese, which I accepted gratefully, eating slowly in the dark, as he returned to crack black pepper over it, a narcotically gregarious maître d', as the burning wood, a final gift from Gaia, popped and cracked itself, punctuating the roar of her battalion of shrieking bugs.

In the velvety tail-end of the shrooms, the three of us sat around the dancing flames, reliving old adventures and trying to fit together paradoxical jigsaw pieces of the universe. I made everyone a steaming mug of sweet milky chai, then headed off to my tent, and fell asleep listening to the trill falsetto of shrieking bugs, reliving the day's wonder as I breaststroked slowly down the astral plughole into a realm where light swam up to meet me.

CHAPTER 21

NATIONAL SERVICE

I returned home late the next day, spending the rest of the week in some lightly phantasmic but still lucid daze. I found I could suddenly eat a lot of things I hadn't been able to for ages, and my debilitating gut pain had vanished into nothing. I thanked the Mushroom Gods daily for their hyperdimensional surgery and returned to work renewed – peaceful, healthy and fascinated by the swirling energies pouring out of my customers. The shrooms had birthed an awareness in me beyond the five known senses. It was faint, but it was unmistakably there, to the point where I swore I could *feel* what customers were feeling as they and their energies poured in through the parting glass.

At the same time, I got the best possible birthday present from my sister, who thought I needed some more love in my life. Madeleine was an eight-week-old French bulldog puppy with a tiny pinch of pug. Frenchies are a confounding pastiche. Equal parts goof, charm and rambunctious fur-clown. Just looking at them is a comedic exercise of duelling improbabilities: flat faces, goggle eyes, mushed-in nose and upturned ears make them look like a fruit bat spliced with a 1930s biplane pilot, dipped in liberal amounts of Gucci. There are not many creatures on this earth that have been bashed with the ugly stick yet remain undeniably adorable, but Frenchies pull it off with trademark elan.

Frenchies do everything with everything they have, gloriously unburdened by inhibitions, temperance or common sense. Their personalities impossibly dwarf them, often to the point of performance art.

Madeline immediately took to my bed, booting out a decidedly unimpressed Moby and worming her way under the covers to snuggle against my side, the hottest hot water bottle imaginable. Frenchies snore like a small warthog with emphysema. Your nights (or in my case, days) will become symphonies of snorts, gargles and impressively pungent farts. I've often wondered how such a small package can emit such heroic amounts of methane.

Maddie spent her first weeks running frenzied laps of the house, bug-eyed and boisterous as she barrelled into the horrified cat, stopping only to ravage one of my shoes, jerking it into the air but getting thrown back with its weight, landing in a confused but deliriously happy heap.

I woke one afternoon, bleary-eyed and haggard as I stumbled from my room into her newest trail of destruction. Several of my favourite CDs had been removed from their places and were now shards of plastic and cardboard, which were beginning to merge with scraps of paper covered in text that were littered across the floor. As I headed into the kitchen their number increased, accompanied by coloured bits of cardboard that were starting to become disconcertingly familiar. Like a distraught Hansel, I followed the trail out through the newly installed doggy door, where she lay mockingly, working on the remains of my limited-edition Kerouac with her back teeth.

I was distraught, but I'd already learnt that Frenchies were fiercely sensitive souls. While they are cunning little felons, they take even a smidgeon of criticism as a personal affront to the soft fleshy part of them that just wants to love you, deflating instantly, then moping around the house for hours until you ladle out a small dollop of love, instantly coming alive with a joy so full and complete that it never fails to lift you. I decided that instead of discipline, it was time for her first ever walk.

It was pre-dusk and Madeleine – lead placed helpfully in her jaws – was leading me excitedly out into the dying day, her young wet nose contorting at the endless olfactory wonders swimming through the air. Everything was coated with a yolk that glazed the colours into a dream as the world morphed into a Caravaggio.

And to be honest, I felt kind of the same way. My connection to the Mushroom Gods had lessened slightly, but a significant portion of *something* continued to linger. Whatever it was, it gave me the most vividly lucid dreams of my life and, if I worked at it, held me connected to something vast, loving and omniscient. As the days warmed and we headed toward Christmas, I ambled about, not doing a lot.

The whole country had Kevin07-fever, as Rudd swept to power, but I was more given to whimsy, interested in shoe-gazing and spiritual pursuits, mostly meditating and placing my bare feet in the earth to commune with Mother Nature. I spent weeks drifting through inner temples, soaking up their narcotic peace, and as I did, I felt the pull to create again. After I finally graduated uni, I started writing again.

Poetry, mostly, that I submitted to numerous literary mags. But something kept drawing me back to my hundreds of fading register receipt rolls I'd scribbled on maniacally over the years between paralytic drunks and impatient taxi drivers. A little voice in some heretofore undiscovered chamber of my brain was telling me that maybe, just maybe, there was some addled gonzo tale that could somehow be stitched together from this elongated cornucopia of smooth-brained Gumbletons.

As I read back over them, I realised no one in their right mind would actually believe them all, but then I'd head into work that night and soon reassess, returning with yet more crinkling pocketfuls of thermal paper snakes covered in rapid-fire ink.

Unable to get the night off, 2008 dawned with me standing beleaguered on an empty petrol-stained forecourt, listening to the far-off shouts and crackling fireworks, thinking of Julia, of ocean sunsets, wine, dark chocolate and glorious stoned sex amid roving clouds of Omani frankincense. But those dreams were soon shattered as the sloshed hordes forced me back to my cage as they came pouring in for supplies, with all their spew, slurred laughter and festive belligerence.

While home life was a carefree meditation of warm, well-beered afternoons with Stevo, or barbeques at Diablo's, in my bright little box, all the hucksters, hoodlums and heroes were getting to me again as they marched in, pilfering, smashing, grifting and grabbing, so in my new state of literary creation, I disappeared into a mental Eden and instead imagined the *ideal* customer; an amalgamation of all the inherently superior qualities of the very best customers, enmeshed with a reciprocal embodiment of all the main pillars of Servo Karma.

This individual would, in the most Elysium-soaked corner of my dreams, enter the forecourt of the service station at a legal, safe speed as their car rolled to a measured halt, with their petrol outlet lined up directly with the pumps, their vehicle not more than 90 centimetres away from the bowser. This person would then turn their engine off (and lights if required) before quickly exiting the vehicle and approaching the pump, all before I knew they were there.

They'd pick up the appropriate pump (the one they actually intended to use), having already pressed the button inside their car to reveal the outlet in which to place it. Then like a skilled surgeon they'd make their incision and get to work, pumping quickly, and before I knew it, *boom* – finished. Pay attention to how, after the deed is done, Mr or Mrs Perfect replaces the pump, screws up their petrol cap, shuts the lid and then strides purposely toward the glass doors. Please note that they do not re-enter their car to forage around for discount dockets or small change. They *may* move their car away so the person behind them can then fill up, but in general they are on a single benevolent mission: to make my job just that little bit easier for 180 blissful seconds.

They will appear through the opening glass doors at precisely the same time I approach the register from whatever menial task I was undertaking, armed with only a smile and a bank card, while announcing the number of their pump in a clear steady voice. Please pay particular attention to the fact that they possess no bag of any description that would enable them to ferret around in aimlessly, only to give me coins or the wrong card.

Nor do they turn on their heels, marching off in a last-minute remembrance of forgotten milk, bread, chocolate or newsprint.

We may exchange pleasantries, perhaps make a perfunctory comment on the weather, but they then sincerely thank me for my service, and I for their custom, then they are gone, like a dream, no doubt off to brighten the lives of console operators everywhere; a veritable Santa for the poor souls who toil endlessly in those brightly lit trenches.

One thing Mr or Mrs Perfect would never do in our three-fold utopian dreamland is ask to use the outside toilet.

In six long years, I never set foot in it. Not once. It started out as a bit of a joke about the cleanliness of the place and slowly turned into me realising that I never wanted to go in there, not even for the sake of curiosity. Others, though, seemed to treasure this Bog of Eternal Stench. For taxi drivers the outside toilet was their haven; four square metres of squalid peace where they could, as one poet put it, 'free the legless brown dog to the sea'.

At this point, I feel compelled to educate you on some servo dunny vernacular. Cabbies weren't the only ones who had some great names for hanging a bog. It seems Stevo's 'snapping off a grogan' was only the tip of the long brown iceberg, so to speak. This quintessentially Aussie cultural movement to invent more and more creative ways of saying 'taking a dump' had spread far and wide, just like your butt-cheeks when you – sorry, I'll stop.

Most of these were from sardonic Aussie tradies. All those portaloos on building sites bred an army of wordsmiths well-versed in 'riding the porcelain pony'. Some other gems were:

'moulding an action figure', 'birthing a politician', 'squeezing a fresh slurpee', 'building a log cabin', 'visiting the chamber of commerce' and 'stocking the pond with some brown trout'.

They had, of course, a different set of names for diarrhoea: 'riding the gravy train', 'the human espresso machine', 'jalapeno tugboat ride', 'supersonic sewer sauce', 'the green apple splatters' and simply, 'the ten-twenties'. I had to ask the guy what he meant by ten-twenties, and he replied, 'Ten seconds to spurt it out, and twenty minutes stuck wiping your arse.'

Some used the toilet for other pursuits. One bold fellow grabbed some 'reading material' from the newsstand as he rushed out the door with the toilet key. 'I'll bring it back. I promise!' he called over his shoulder. But most people, whether on an ice binge or just having a bad day, simply favoured grotesque wanton destruction of that poor little cube.

After another morning spent cleaning the blood and shit off the walls, this time accompanied by several syringes floating in urine from the uprooted toilet, Barb had, after getting it fixed and cleaned up for the last time, made the final decree in the diary, thrice underlined in red permanent texta:

WE *NO LONGER* PROVIDE A TOILET FOR CUSTOMERS!
NO EXCEPTIONS! IF ANYONE DISOBEYS THIS
THEY ARE IN DEEP ... SHIT!!

One by one the customers reacted with disbelief when I told them the toilet was no more. Taxi drivers stared at me with venom pulsing in their eyes, as though I was the one who took away their haven. All in all, though, after a week or so, people eventually began to accept the new reality.

Early one morning, just as the sky was turning from black to charcoal, a yellow Lancer screamed in, weaving between the pumps and jerking to a halt just before slamming into the gas bottles. A man jumped out, rattling his shoulder against the glass door I unlocked just before he ploughed through it. He was breathless.

'Can . . . I *please* have the . . . key to your toilet?'

'Sorry mate, it's out of order.'

'. . . What's wrong with it?'

'Well, nothing exactly, I just can't let people use it anymore.'

'Could you make an exception? I *really* need to use the toilet.'

I was about to say no but something about the guy swayed me. His sallow cheeks were contorted into a look of pain, while his lips were so dry it looked as though they were covered in spider webs. He had a sheen of sweat glowing on his forehead.

'Okay,' I said, offering the key, 'just don't make a mess, or my boss'll kill me.'

He snatched it off me and sprinted around the corner as I went to stock the fridge.

The fans that cool walk-in-fridges have a loud hum. You can be in there stocking bottles of V while a riot was going on outside and be none the wiser. But I heard the screams. They cut through the white-noise of the fans as clear as night.

As I rushed out of the fridge, they became even louder, and at first I thought I'd forgotten to lock the door and that some tortured lunatic had made his way inside. It took me a while to realise that the screams were coming from outside, from the toilet.

They continued, on and off, for eternity, as all sounds of horror seem to. Eventually they stopped and after another five minutes the poor guy came back, the sweat on his forehead now in beads. His demeanour had taken a turn for the better: he was no longer tensed up, in fact he seemed deflated, like he'd just run a triathlon.

'Kidney stones,' he breathed. 'Sorry, I locked the key in there. Too . . . much . . . to . . .'

He staggered back out to his car, clawed open the rear door and fell in: a worthy candidate for the last ever person to use our toilet.

One of the best things about working early mornings was hearing the hilariously creative conversation between tradies not bothered by my inescapable eavesdropping:

'Hey knackers, gimme three bucks for this pie.'

'Fuck me dead, tits on a bull you are. Get a job.'

'Give me one, you pole-smoking bandicoot rapist.'

'Shove it up your cloaca, you insidious little rat-clown.'

'What the hell's a cloaca?'

'The cloaca's a hole for birds. They shit out of it, they fuck with it, and the females lay eggs out of it.'

'. . . Good insult.'

'Yeah, your mum thought so.'

'I'd suck a fart out of your mum's arse like a bong hit.'

'I'd walk a mile over broken glass just to hear yours piss in a jam tin.'

The ladies, though, gave them a run for their money.

One November morning, I had a shitty cold and the tail-end of a painful chest infection, but knowing I'd get no mercy from my area manager, I'd just shown up, dosed to the gills on cold and flu capsules. I was pretending to stock the Eclipse mints and chewie, but was actually listening to two tipsy young women drinking their vodka-Red Bull-slurpees while waiting for their taxi.

'What the hell is her problem with you anyway?' asked the redhead.

'She's just mad 'cos a few months ago we went through all her Facebook pics and tagged her crotch as all the guys she's fucked,' replied the blonde.

'Seriously!? That's gold, Rach, fucking *gold*!'

'That, and I kinda fucked her ex-boyfriend.'

'Wellllll, that'd do it . . . Hang on, was that Rob?'

'Yeah, do you know him?'

'Not really, but I know his little brother, Tom, my mum used to babysit him when he was younger . . . Actually, that guy last night kind of reminded me of him a little bit.'

'Seriously? He wasn't that bad.'

'Did he give you head?' asked the redhead, not bothering to adjust her voice to a more surreptitious level.

'He used a semicolon in his text, Jane, of course he wasn't gonna go down on me,' said Rach.

'Hah! Really?'

'Yeah. I get the feeling he probably went home and told his mum about it. I swear, at one point last night, his moaning reminded me of Gilbert Grape.'

At this, Jane couldn't hold it in anymore. The slurpee she'd been drinking exploded out of her mouth and onto the already grimy tiles. As she wiped her face, still doubled over in laughter,

Rach, giggling herself now, continued, knowing Jane was on the mat and begging to be finished off.

'It felt like I was being body humped by a twelve-year-old.'

'Stop! My sto – my stomach hurts!' It was that kind of still drunk, sleep-deprived laughter I knew all too well, the kind that just doesn't quit, taking only one well-timed comment to set you off. Jane fell on her arse, dropping her slurpee to explode bright goo all over the floor, switching between cackling and trying to get oxygen into her lungs.

'And what about you and Fabio last night?' Rach probed between giggles. 'What kind of man walks around in *silver tights*, for fuck's sake?'

'STAHP!'

'And what was with his facial hair? He looked like he just ate out someone's ass.'

'PLEASE, I'm b . . . begging you!'

Eventually, Rach stopped with the questions and Jane arduously made her way back onto her feet as the laughter between them faded to sporadic giggles.

'Fuck, my stomach won't stop growling!' said Jane.

'Probably because there's nothing but semen in it,' offered Rach.

Primed like explosives, they both erupted again. And soon enough they'd infected me too, as painful as it was. They turned and saw me laughing, like some broken animal by the side of the road, which of course made them laugh all the harder. We laughed until we couldn't breathe at all, as customers came in mystified at what was so devastatingly funny, and then they started laughing too, until big whooshing gasps were all that escaped from our mouths and these, of course, only made us laugh even more.

While the multifaceted orifices of avian creatures or the virulent alchemy of uncontrollable laughter can present as enduring mysteries, I've discovered you can tell a lot about a person by the way they treat a retail worker. I've found that, generally, if they bother to offer a greeting and actually engage their facial muscles in some fashion, then they're probably an okay person. The thing is, though, so many people consider this an inconvenience too taxing to overcome, simply preferring to grunt, or, as most of them do, simply stare straight ahead and hope I'll shut my pie hole. I suppose I should at least appreciate them being honest.

But that's the problem, you see, dearest customer: the whole *thing* is an illusion. I'll be honest here: while meditation, psychedelics and other Eastern practices have calmed me somewhat, I've found that even *their* purifying strength seems to quail before the onrushing tsunami of stupidity, insolence and condescension that washes over me nightly. Before I started, I was a patient, mostly trusting person who felt that, generally, people would give you the benefit of the doubt and act, well, pretty much like respectable human beings. Many years later my beliefs have changed.

You see, people are dirt. At least when they're in servos. I have this theory about the way they believe they're entitled to act. I still don't have a name for it. It's like a hybrid between road rage and customer-is-always-rightism, taking the worst parts from both and melding them into a new super-powerful excuse to act like a bell-end. *I am customer, hear me roar!* Ask any service station attendant and they'll agree wholeheartedly.

Knowing they'll get away with it, customers will do and say things in a servo that they wouldn't even consider doing in a supermarket, KFC or Macca's.

Therefore, dear customer, I don't like you. I never have and I never will. Whenever you drive in and grab a pump, a machine next to my head starts beeping like a shrieking harlot and in my confusion, I look up to see you glaring at me, replete with your impatience and haughty look of entitlement.

And to compound this, you're getting ruder. Every fricken day. I don't give two shits about how you are, but I *pretend* I do, if only so some of you will remove that sullen look from your faces and stop treating me like some robot serf. I will then reciprocate in kind and the place will suddenly be filled with goodwill that is about as genuine as the chicken in our burgers, but at the same time useful in lubricating the inveterately soul-destroying machinations of the customer production line, getting you out the door and me back to my soft porn and sausage roll, both with smiles on our faces.

Therefore, those who know intimately what's it's like in today's retail trenches, who stare pricks in the face all day, soak up their rudeness and vitriol and get paid $28 an hour, are not the problem here.

These days, when 'war' equates to your smart TV showing distant green fireworks on CNN, and 'patriotism' to having a cold beer in a wading pool on Australia Day while listening to Triple J's Hottest 100 – the idea of conscription is not one that currently sits at the forefront of the nation's psyche. In this instantaneous age of TikTok and Snapchat, the people of today fight not for their country, their valour or some age-old idea of 'rights'. What we want is stuff. Our battle is a material one.

With this in mind, I suggest we bring back national service. Two years' worth for every able-bodied human over the age of fifteen. But these soldiers won't be facing the bullets and bombs of opposing nations. They will instead be facing hordes of rude, angry and impatient customers.

That's right. I reckon everyone, especially those over thirty-five, should do two years of retail service. Just like the army, you'll learn how to lead under pressure, meet new people and develop new skills. We'll even pay you to learn (Cert. III in Retail Operations only, sorry).

But most of all, you'll return to civilisation with a renewed respect for the men and women who patrol the nation's retail frontline daily, or nightly, as some poor souls must endure.

Come and help your nation on its path to a glorious future, one we can all share in – where even the biggest arsehat on earth can remember their fucking pump number.

CHAPTER 22

THE MOMENT™

It was 3.46 am and on two hours sleep I was serving a bespectacled tightarse who wrote all his purchases down in a little red book. He had, after much deliberation, relinquished a $50 note and his oldest discount docket, rearranging the others like the world's saddest deck of cards before returning them to his wallet. I looked at him like a man peering through a fish tank, only to ask, 'Got a discount docket, mate?', as I swiped the folded $50 note through the card reader.

I'd stopped meditating again. Although I needed it more than ever, it had become harder and harder to find both the time and the quiet in which to seek out peace. Thus, the dark circus of anger and insecurities came marching back into my mind like some raucous tin pot parade, my ego at the rear, dressed in full royal regalia, cockily blasting his infernal trumpet. I was firmly back on the not-so-merry-go-round. And this time it was speeding up.

Customers were getting worse. Ruder. Stupider. More self-assured in their incompetence and entitlement. Snooty people who'd ignore my cordial greeting and ask for rare brand-name cigarettes, in Werribee, of all places, looking down their noses at me as if their farts smelled of newly opened tennis balls. Idiots who parked in between two lanes on the wrong side of a bowser and grabbed a hose five metres away, stretching it to breaking point, twisting it upside down and then glaring angrily at me, beeping their horn and throwing their arms in the air when it didn't pump. Geniuses who burst their bike tyres using our air pump designed solely for cars, storming into the store demanding retribution. Meanwhile, I wanted to erect a flashing sign, proclaiming, *Hey Big Spender!* for the spendthrifts who came in to waste my time for 65 fucking cents of petrol. It shouldn't have got to me but it did.

Eight Harleys roared in one night and formed this impenetrable line across the forecourt. They sat, these bikers, took off their helmets and smoked cigars. On the forecourt of a petrol station, where petrol vapours rise like yeast. I left them and went into the fridge, only to be dragged out by the *BEEP BEEP BEEP BEEP* of the pump authorisation. One of the bikies was sitting on his Harley. He'd picked up the hose and placed it in his tank, centimetres away from his groin. Oblivious to the fumes and countless signs advising against such an idea, he sat there impatiently waiting for the petrol to pump. He held his burning cigar in his mouth.

After ten seconds of waiting, he looked up and yelled at me to authorise the pump. Having removed the cigar from his mouth, he waved his hands around, sending hot ashes onto the concrete.

The store was in the midst of a massive refit, inside and out. New tanks, a complete overhaul of the console, all that jazz. Every day was a confused mix of chaos, with workers stationed inside and out, bulldozers and earthmovers duelling with entering cars outside and workmen using tile saws inside. Some will know this aural pain and suffering all too well. It's an excruciatingly loud, shrill noise, and when it's happening, thought itself is impossible. All the tile dust, plaster and debris hung in the air, choking all those in the vicinity.

But such hazards were nothing in the face of rapacious predatory capitalism.

We were given masks and – in between screaming tile saws – told to stop whingeing. While this was great early practice for Covid, imagine walking into a dusty, chemical fume–filled servo to be assaulted by the piercing scream of a tile saw and a creature with pale white skin, dusty clothes, an N-95 mask on his face and a vacant stare. That muffled noise from beneath his mask is him asking if you wish to take advantage of our great specials.

To make the situation even more entertaining, the workers blocked off our exit. Cars would venture in over the mountain of rubble, drive around for three minutes trying to find a working pump that didn't belong to a dug-up tank and finally fill up with a scowl, glaring at us in preparation for the tirade they would release when they made their way inside. But once they did, they were drowned out by hammers, nail guns and the world-ending shriek of the tile saw, soon storming back to their car to speed toward the blocked exit, only to screech to a halt in the rubble, then drive through an obstacle course of earth movers and moulded plastic gates to exit at the obscure back entrance to the store, almost wiping out every entering car as they did so. And then the cycle would begin anew.

Heaps of regulars had left at this stage, so amidst the rubble, I was surrounded by newbies, marvelling over their variances of $2000 a night. Every second night seemed to be a 'buddy shift', where I was supposed to train a hapless new graveyarder in the dos and don'ts of servo nightlife. I stressed the point to Barb that in my current condition she probably didn't want me corrupting such fragile young minds, but there was no one else to do it. They'd show up bright and early in their sweet little Diahatsus and Festivas, and be waiting for me when I walked in on the dot of midnight, haggard and dehydrated. I'd leer at them and order them to fetch me a can of V from the fridge.

Eager to please and full of questions, they followed every procedure they'd learnt from training vastly more extensive than mine to the absolute letter and marched around the store with a proud, purposeful rhythm. They'd bound up and ask, 'How does this work, Mr David?' The loss of 'me time' was bad enough, but endlessly talkative Burmese students, all bright-eyed and bushy-tailed at 3 am, was the icing on a shitty cake.

But things always sweetened when they got their first taste of chaos and froze, like a deer pinned by the headlights of a groaning semi. I'd let the freaks play with them a bit; give them a glimpse into their not-so-distant future and see if they could hack it. On his first ever Saturday night, my first young tyro, Aravinda, was accosted by a mangy servo-to-servo salesman who was intent on selling him several pairs of lacy women's underwear from the dozens concealed in his garbage bag. He tried to say no, but bogan criminal salesmen have an advantage over their employed counterparts: a complete and utter lack of mercy or common decency, coupled with a fiscal determination that would make hedge fund managers blush.

'C'mon mate, you got a missus waitin' for ya at home? Buy her summa these 'n' she'll jump ya fucken bones!'

Aravinda shook his head rapidly.

'No missus, eh? Well, they're pretty comfy either way!' He lifted up his jumper and pulled down his pants slightly to show Aravinda how much he believed in his product.

Recognising the teachable moment for what it was, I ignored my student's desperate eyes and announced I was off to stock the fridge, in the hope the boy could now become a man. Ten minutes later I emerged to see him with a look of relief on his face.

'Did you get rid of him?' I asked with a half-smile.

'Yes, I told him to go and get out,' he said proudly, looking to me for reassurance.

'That's what you've gotta do sometimes, unfortunately,' I mused, as I noticed a splash of pink satin in the console bin.

But such snippets of joy were becoming increasingly rare. To get rid of the ancient stock from shelves we no longer needed, we had a fire sale of all the useless crap nobody ever bought. We had this massive bargain bin in the middle of the store and it attracted every grifter under the moon with the power of a Venus Bogan Trap. They'd come in at 3 am, smack-bang in the middle of my quiet time, and spend twenty-five minutes sifting through the obscure items, continually screeching the same question at me while I was trying to read: 'Are ya *sure* it costs this much, champ? That's too cheap, it can't be right . . .'

After my fourth affirmative urging they'd approach the counter with armfuls of booty: rolls of electrical tape, bags of rubber bands, torches, jumbo packets of baking soda and exotic, failed chocolate bars. The whole lot would cost about $3. They'd then contact more bogans who'd come in guffawing and repeat the entire process with the same sceptical questions, ensuring my nights had no peace.

Ralph was also ensuring my nights had no peace. With no real life to speak of, he often seemed to struggle to find things to spend his weekly wage on. Being a Gemini, he decided to devote a not insignificant portion of his weekly wages to expressing himself via commercial expenditure. Every month or two, he'd splash out on a new gadget, quickly learning all its features so he could then ceaselessly annoy us all with the damn thing.

I'd been at work only a few minutes when Ralph pulled out his new phone, taking great pride in the countless ring tones he had already loaded onto it, playing them for me, one by one.

'This one's "Put Your Hands Up For Detroit",' he said, following me around the console, holding the phone at the optimal angle, just to make sure I clearly soaked in every possible decibel. I tried to dodge him, and like two opposing magnets we continued our merry dance as I visualised inside my mind the things I would do to him if the pillars of law and order happened to rapidly collapse and the world suddenly possessed no consequences.

'This one's called "Everybody Dance Now",' he yelled over the screeching din as a stern pudgy man walked toward the console with a look of distaste. His discount docket was out of date, and as Ralph's noise pollution droned on, he kept staring at this crumpled bit of paper, as if willing it to change. He also had a little book with him to write down every purchase and he darkened when I told him I couldn't accept his docket.

'That's going to throw out my whole budget for the week!' he huffed, with the look of a petulant child. He gave me the silent treatment for the rest of the transaction, which was actually anything but, with Ralph's distorted melodies knifing through the air and into the soft white meat of my brain. I watched the tightarse stomp back across the forecourt – seething over 73 cents – toward his $80,000 4WD, thinking he probably ripped 'December' out of his calendar so he didn't have to buy Christmas presents for his kids.

As 'Eye of the Tiger's distorted guitar riff pierced the air, I stomped off myself to grab a Dare iced coffee, deposit a double shot in it from the coffee machine, and stock the blessedly quiet fridge, wondering why I was still in this forsaken cube of colour and madness.

Money, mostly. I was still saving and mostly behaving. Watching Maddie grow into an even bigger boisterous and loveable terror. Looking with distaste over crappy, entry-level marketing jobs. And writing, when I could, trying to fit my zany jigsaw puzzle of years of servo madness into some intelligible narrative. Either way I was still fully enclosed within the servo's clammy neon grasp. While it had calmed down a bit (I'm guessing the portal may have finally closed), it still sometimes had resurgences that brought me reminders of what it was like in earlier years. Freak City. Currently stuck in the middle of one of these renaissances, I felt I was in a cereal bowl full of fruit loops; some crunchy and dangerous, others soggy and defeated. But I wasn't the spoon; I was the milk: becoming tinted a thousand different shades of madness as the anarchy of Gumbletania began to once more seep its way into my pores.

As had happened before, a single night really made me give some thought to my life choices as I reached an absolute apogee of Gumbleton madness. It was 1 am. I was haggard, sallow and the downlights of the canopy were sour milk in my crusty eyes. I was standing outside by the gas bottles, working my way through a king-sized Red Bull and a king-sized Cherry Ripe, full of contempt for the coming morning and all the cars, beeping and rude impatience it would spew forth. As the payload of sugar began to spread its dark alchemy through my cells, I saw something approaching the forecourt at breakneck speed. A young man clad in both lycra and leather was speeding toward me on a bike, proclaiming himself as 'Big Kev'. He was, as he had no qualms about expressing, 'excited'. Quite excited.

He rode past me into the store and started pedalling down the aisles as I rushed back inside. He had a garbage bag – painted with white stars in the design of the Southern Cross – fitted loosely around his neck, which I took to be some sort of cape. He punctuated his patriotic jargon every now and then with explosive operatic singing.

After five minutes of this, Big Kev departed, only to be followed into the store by a psycho with a needle half-filled with a light-brown liquid, rasping, 'I'll swap ya this for a carton of cigs! C'mon *maaaate*, it's the GOOD SHIT!'

He was followed by another psycho, this one holding an ice pipe, who exited his Datsun 120Y and then stood dead in the middle of the opening auto-doors where the sensors didn't reach, so that they'd open then close slowly on either temple of his bald shiny head as he sparked the lighter and grinned at me.

Then, at 3 am, in the dead of night, an LPG bowser started spewing out thick clouds of gas, soon filling the forecourt.

Pressing the 'stop all pumps' button did nothing, so I rang triple zero. The firies came screaming in inside five minutes, but were unsure about how to stop it, so three of them stood over the gas bloom with fire hoses at the ready, costing the servo $1600 every fifteen minutes. In the meantime, Werribee's night-faring population just ignored the anxious firefighters and their several chemical fire hoses aimed at a hissing gas pump and drove right on in, happily walking through the valley in the shadow of death for Cheezels, Picnic bars and veggie pasties.

My personal favourite was the guy who drove up to the entrance, only to encounter the big yellow bin I'd placed there, surrounded by emergency tape blocking the driveway. He sat in his car for five minutes, occasionally beeping his horn, then started creeping forward, pushing the bin into the servo with the nose of his car. He took a wide berth around the two fire engines with their flashing beacons, cruising past the throng of fully decked out firefighters staring at him with pure disbelief, and came to a rest right outside the glass doors. Strolling in, he ignored my look of horror and asked for a packet of Winnie Blues. 'And a lighter, fanx champ.'

Ten minutes later, Randall walked across the forecourt with his guitar slung over his shoulder. Knowing his habits, I launched out of the console and tore outside, yelling '*No!*', just as he'd placed a smoke in his mouth and was reaching into his pocket for the lighter.

In the past I would have been shaken up, or at least would have shaken my head in disbelief. These days they were just interruptions, annoyances. Votive offerings to the shrine of beguiling stupidity. All I wanted was a sausage roll and a quiet sit down, but the freaks wouldn't leave me in peace. So I began to care less than ever before, smothering my anger with a tired drollness. I'd watch the forecourt brawls, the drug arrests and the vile threats spat at me with detached disinterest. I'd find stock that was a year and a half out of date and laugh. But it was never enough. There was always something to piss you off, something to cut through your tired disappointment and fire up the rage that lingered inside of you.

Like a guy covered in soot refusing to pay for his petrol because the pumps were 'too dirty', yelling insults at me as he retraced his black footprints back to his decrepit ute. Or when we received our new point-of-sale equipment and customers would continually press the cancel button, thinking it was the credit button, yelling at me repeatedly when it didn't work, two thousand times a week.

Around this time, Ralph really started hitting the clubs. Well, the *club*. On the rare occasions my mates managed to drag me there, I'd see him on the dancefloor or circling the upstairs bar, paralytic, even though the night was still in its infancy. To his cricketing buddies, he was the main attraction: a man-child who was almost thirty but couldn't stomach more than three drinks. And when those drinks were Smirnoff Blacks, then he could manage to get down eight before he started to think twice.

One by one they'd drag him to the bar, an arm draped around his unsteady shoulders as they pumped more and more alcohol into him, cheering on his slobbering deterioration like thugs watching a cockfight. One of them came up to me at one stage and bleated in my ear, 'Ralph couldn't get a root in a brothel with a hundred bucks stapled to his forehead – he's tried for seven sheilas tonight and got rejected *eight times*!'

As he brayed like a donkey, another of them was pushing a hazy Ralph into the path of a curvy blonde, filling him with false hope and then revving themselves up for the entertainment to come. I couldn't stand it. The poor guy was finally getting the adolescence he so greatly needed and it was being turned into a freak show by these vindictive, merciless little cretins who only wanted him around to torment.

As I stood there watching them, the lasers bouncing off their cruel faces, I was reminded of a book I saw poking out of Andrzej's bag one night: Ralph dredged up strong memories of Piggy in *Lord of the Flies*, the poor vulnerable boy set upon by his vicious peers. I suppose it was his rite of passage, but it still saddened me.

Little did I know at the time, Ralph was doing it tough. As the winter of 2008 finished and the customer numbers began to grow, he was seriously struggling. Angry, disjointed and unresponsive, he was beginning to worry me. I wanted the old Ralph back, but it seemed he had been spirited away, leaving only this darkened shadow.

One night I came into the servo to fill up on his shift and, during a lull between customers, we had a real D&M. He said his life was slipping away behind the console. I said I knew the feeling.

'I see all these people come in every night and they're smiling,' he said, 'not because their company tells them to, but because they're actually happy. Guys with girlfriends . . . I've never even kissed a girl. I'm twenty-nine . . . I haven't told anyone before, but I've suffered pretty bad depression because of this job. I mean, I don't really think I'd do anything stupid, but . . . yeah . . .'

He trailed off as what he said sunk into me. I thought about my own feelings over the years. All the insanity this place seemed to breed inside me. A feeling that all my dreams were made. Chained to the servo and its dull charades. It definitely wasn't healthy.

'Have you thought about what you can do?'

'What *can* I do?'

'Quit.'

The answer crossed his face like the light of a new day. But then just as quickly he rid himself of such foolish dreams. 'But what else would I do?'

'Who cares? Anything. If that's the kind of effect this place has on you then you've gotta think about it. Seriously.'

'Yeah, maybe I will,' he muttered, as an angry truck driver caught his eyes, lifting his arms in the air, questioning why the diesel pump wasn't working.

Anyway, as I found out the following week, the reason our store got rid of so much stock was to make room for the newest member of our family: the granita machine. With eight different flavours, it was futuristically styled but at the same time looked like it belonged in a 1950s diner. The lids to each granita flavour would open from the top, so I spent many nights tempted to tip a bottle of cheap vodka in, Stevo-style, and see what happened.

Because everyone simply *had* to have all eight flavours in one cup, they would always overfill the bloody things that never quite closed properly and leave a rainbow of putrid sugar stuck to the floor and counter. One knob-end would bring four overfilled granita cups to the counter and would be followed immediately by someone resting his clean white-shirted sleeves in it and then abusing me for it. It got worse as summer approached. If we didn't have time to mop every half-hour, it stuck to the floor like shit to a blanket and soon ended up looking like one of those crappy community art murals done by kids which are stuck up on school walls and on the sides of underpasses.

Sometimes I'd come in and the floor would have an actual texture along with the colours. I had to cut through this with a mop nightly, eventually resorting to scraping the dried granita off with whatever was sharp enough. It sometimes took an hour, always punctuated by idiots coming in and covering the floor with more sugary ice, even though the damn machine was off and covered with 'out of order' signs.

One night, I'd had forty minutes of this, and after finally getting the floor clean, I returned to the console ready to snap. I could feel it building when a VN drove slowly across the fore-court and launched several bottles out of its windows, coating a portion of the forecourt with jagged fangs of glass. These were promptly driven over by a woman in another VN, who, after hearing the signature crunch under her tyres, barged into the store and started blasting me, powering up the rusty chainsaw of her voice for its nightly lashing. I stared at her while she roared away, zoning out for a second or two. Not listening, just observing, watching the spittle fly from her lips and onto the promo packets of green and blue Extra chewie as she rained down hell.

The night continued on, like the squalid circus of old. As I stood there at 3.33 am, trying to comfort a rat-like drunken girl who had discovered her boyfriend's long-standing infidelity and listening to a shrivelled old man complain about the price of a peppermint Freddo Frog, I experienced The Moment™. I felt this massive red surge of futility rising up through my body. It rushed to my head, pulsing with an angry purposeful rhythm.

Again, I stayed completely still as the elderly man continued to seethe, zoning out and focusing simply on the huge Gorbachev-like birthmark splotched across his forehead. As I bored my eyes deep into this crimson blob, I felt this massive universal clock ticking somewhere; time was flying by and I was marooned in this godforsaken place, this other twisted dimension, where only the damned remained and time had not just stopped, but, bleached away by the brutal downlights, had ceased to exist at all.

I don't know how long I stared at him for, but it was a while. My mind was still pulsing at me, asking me why, just like Ralph, I had wasted five years of my twenties, of endless weekend nights, in this lurid suburban hellhole, and why I didn't have a nice normal job where beaming grandpas came in, ruffling the hair of young children with a warm smile and a Werther's Original. I wondered what the hell this old bloke was doing up at three in the morning anyway, and then a car horn sounded, shattering my trance. I cracked. I picked up a blue Eclipse tin and pelted it at him, aiming to hit the angry birthmark that had by now become bright scarlet. It missed, but it shut him up. I took a deep breath and gave him an option, in a steady controlled voice that wasn't quite my own:

'I am going to quit tomorrow, old man, so what happens next doesn't bother me in the slightest. Take the Freddo Frog. In fact, take the whole box. But get out of my store in the next ten seconds or I will pick up this metal pipe I have under the counter and beat you with it, you bitter, shrunken fuck.'

I knew then that nights and I had come to an end.

That morning, after my cover had shown up and I'd counted my till and done my paperwork, I stayed an hour after my shift had finished, and when Barb bustled into her office, I swivelled around on her chair to face her, stalker-style.

'What's going on, Trouble?' she said, eyes narrowing on me.

'Can we have a chat?'

CHAPTER 23

DISCONTINUED PRODUCT

My last ever weekend of graveyards was one long requiem, but with an undercurrent of joy pulsing just beneath the surface. I was like so many others I'd worked with: they'd have this lambent glow in their eyes to rival the falling bath of halogen as they strolled the store, smiling munificent goodbyes at bemused strangers. You could *feel* the anticipation of freedom that oozed from every pore of their being as their handcuffs began to finally unshackle; an inmate emerging from their dingy cellblock, reclaiming their personal effects, then out through the opening gates and into the purer light of the sun.

To celebrate, Andrzej had suggested bringing in a ridiculously strong bottle of home-brewed hooch called Palinka, which he assured me was distilled to a proof that made it illegal in its country of origin. His anarchist soul wanted to pour it into the pineapple-flavoured granita and let the chips fall where they may. The main problem among many with such creatively destructive bridge-burning was that we'd have to use up all the Palinka granita ice by morning, which he assured me meant that even all that pre-diabetic truckload of icy sugar still wouldn't keep us on our feet. I settled for a bottle or three of Kahlúa, which I dutifully spiked into multiple Big Ms and coffees over those joyous final nights.

I spent most of it saying goodbye to all the freaks I'd shared my time with over the past five years. Yet so many were gone. There was a different bevy of drug dealers by now, the old ones in prison or God knows where. Most goodbyes, when they came at all, were perfunctory, but I didn't care – I was sozzled on Kahlúa, yes, but that light at the end of my tunnel, that tiny coin of sun at the top of my grimy well, had grown blinding with its unimaginable proximity.

The final night, with my last ever delivery of stock, I felt a weight blessedly removed from me, like Atlas finally having lifted that hatefully colossal globe from his shoulders. I looked at it and sighed, not looking forward to the haul but at the same time relishing it, because it was the last fucking one. Ever. Four pallets, each stacked almost to the roof with boxes filled with drinks, chips, chocolate bars and other assorted crap. Probably about a tonne in total weight. Keeping in fashion with the previous five years, it was delivered two hours late. Yet I set upon it like a lion would a fallen impala, feverishly ripping open boxes and carting drinks to the fridge. Sometime just after 5 am, as the sleepy cars were beginning to roll in with the rising sun bouncing off their windscreens, I stacked the final box of 2-litre Cokes that wouldn't fit on the shelf.

It sat there, this box, atop the skyscraper I'd crafted, like the hundreds before it, swaying threateningly. I smiled at it and turned my back, skipping out of the storeroom and back to six idling cars, frenzied beeping, the seven thick bundles of newspapers and magazines, and a scowling taxi driver tapping repeatedly on the glass door with his car keys.

But no. I didn't quit. Because I'm a deluded, masochistic fool, I settled for a lesser version of madness – afternoon shift. Four pm 'til midnight.

Still, compared to my saturnalian nights they were a completely different box of frogs. As I walked from my car to the store to start my first ever arvo shift, I smiled at both the healthy lunch in my stomach and the warmth of the sun on my back. For the first couple of weeks, I was a grinning fool, smiling at customers and silently thanking them, just for being normal. For driving in at a leisurely pace; for not looking through me with dead black eyes; for not stalking around the store for two hours lashing others with sermons from their deluded shell of a mind.

Parents would come in, smiling parents, who'd treat their well-behaved kids to a Milky Way before bedtime. Seven-year-olds would come rushing in with a $50 note and an important look on their face, entrusted with the serious task of paying for Mum's petrol. After being on Planet Zorg for so long, it was like falling into a Disney movie.

I'd jump at any chance to go outside and perform the most menial of tasks: changing the price board; fixing the car wash; emptying the giant decaying bins; and even un-jamming the coins of vengeful, screeching fools, stuck in the vacuum machine.

They'd stand there seething, letting loose an entitled mono-logue on our wilful incompetence, until I simply offered them a $2 coin, gratis, then looked up at the sky and smiled.

Back inside, I'd disperse a line of people in seconds, serving each one with a rapid kind of grace and leaving each with a sincere 'thank you' and 'farewell', encouraging them – unironically – to come back again soon. Some of my old friends from the night came in during daylight hours and halfway through the trans-action, they'd recoil in shock at the transformation: '*Fuck* mate, I didn't notice ya! Ya look, well . . . *fucken normal*!'

Out of work, my reinvention continued, as the glorious sun reintroduced me to the world's brilliant natural colours again. I took Maddie for long rambling walks at sunset by the river, her Frenchie thespian soul leaping her off the embankment to bellyflop two metres down into the swirling currents.

Back home, I lay on a patch of bright green grass in my tiny backyard, listening to uplifting songs on my iPod while writing poetry. Ants and aphids mingled over the paper, weaving crazy circles around my even crazier scribbling pencil. Sparrows chirped and jumped on the back fence like popping corn, awaiting their turn in the birdbath. Up in the sky, currawongs and sparrows leaped from gum to gum, before weaving their way through the trails of powerlines stretched like claw marks slit from the sapphire sky.

Moby came and joined me, lying in perfect contentment as I brushed the grass off his belly and watched clumps of his long hair glisten silver in the sun as the breeze surfed them up over the fence and into the blue beyond.

Every now and then I'd stop writing and close my eyes, feeling the cool carpet of the grass on my back and the warm yolk of the sun on my face as the music rose up through me, lifting me with this rolling surge. I was meditating again, properly this time, for longer and longer. Which meant I was also writing, and little by little, the hundreds of deranged anecdotes were starting to knit themselves into an even stranger technicolour dreamcoat that might one day be an actual story.

U2's 'Beautiful Day' came on shuffle and it lifted me even higher as I lay simply soaking up that moment of absolute perfection when beautiful music and an image merge into a split-second of something special. The trees above my back fence were waving – breathing – in the tiny breeze, their shades of green swimming against the blue, and I felt their energy, their *life*, and as the song climbed to its final chorus and Bono wailed about the colours coming out, I finally understood him.

As the hours of my afternoon shifts flew by, I spent countless minutes watching the sun being slowly pulled to the horizon as invisible fingers tugged on the skirts of the heavens. I'd stand half-serving customers and half-gazing into the deep smoky blue leaking into the golden yolk surrounding the sinking orb, keeping a smile on my face even as the children's shrieks duelled with The Beep for first rights on my sanity, not realising the real me was simply somewhere else.

But reminders were still everywhere. I'd get a powerful sense of empathy seeing other graveyarders succumb to the virus of the night. Andrzej felt it more than most. He'd show up at midnight, shielding his eyes from the hateful sour milk of the lights. He began to function only after draining two and a half large cans of Monster, which for the saner slice of the population is enough caffeine for a one-way ticket to the astral planes.

I looked at him and shuddered, full of compassion, but glad I was never going back.

Late one spring afternoon, as I was basking in the full power of the setting sun, and snacking on Nellie's snow peas and fried goat's cheese, I began smiling as I gazed at a stunning Italian woman pumping Premium into her silver Holden Astra on pump two. Nellie stopped serving for a second and studied me. I turned to face her searching gaze.

'I don't like this.'

'What?' I smiled.

'This new you. I can take the sarcasm, and I can take the whingeing, but this . . . this . . . *positivity*. This I cannot accept, David.'

On one of those brilliant sunny afternoons, I was walking with Stevo from my car to the local bottle shop, when around the corner walked Randall, wizened and decrepit, guitar slung over his shoulder and a look of fierce reproach on his face for the blinding sun.

As I walked toward him, he eyed me suspiciously, but after I took off my sunnies his face lit up, the crinkles smoothing out for just a second as lucidity swept over him.

'Hey David.'

'Hey Randall, how are things?'

'Not bad. When did you quit?'

'I haven't. I'm on afternoon shifts now. I finish at midnight.'

'Oh, okay. Your aura is lighter now.'

'Really? Can you see it?'

'Yes, it's light blue now.'

'What was it before?'

'Green. A dirty, dark green. Good things are coming now. Do I owe you any money?'

'Umm, no, I don't think so.'

'I owe Rathindra money. And some DVDs. Can you tell him I'll pay him back? When is he on next?'

'He quit a few months ago.'

'Oh, that's a shame, he was a good man. It's important to be good. Being good gives the evil ones less power.'

'Agreed, Randall. Are you still going to church?'

'I try to,' he said with a guilty grin.

We talked like this for a while, as Stevo studied the craggy, shrunken man walking beside me, before we left him to head into the bottle-o.

'He's the Gumbleton to end all Gumbletons,' Stevo remarked as we walked past the white wines. 'But somehow, he's like . . .' He paused in front of the Kahlúa and Midori, in deep thought.

'. . . the best kind of Gumbleton there is,' I said proudly.

After I bought my beer, I asked the guy for some PJ 30 Golds, which I presented to Randall on the way out, with a smile and best wishes for the future.

'Thanks David,' he said. 'Stay light blue. It's better for you.'

While I was deliriously happy in my new environment, it didn't mean that I couldn't see how much I was still being exploited. Afternoons made me busier than a vacuum cleaner in a dirt factory, and once I stopped smiling like a loon, I realised that in some ways I'd gone from the frying pan into the fire.

I worked for only twenty-four hours each week at that point, but in each eight-hour shift I was a confused blur. My servo was one of the busiest in Werribee and the next few surrounding suburbs as well. And in all of my twenty-four hours, I had cover with me for only three. And that was for the first hour of my shift. For the remainder, amidst the biblical peak-hour deluge of grumpy homeward-bound steel with empty tanks and callous hearts, I was on my own, serving over 1500 people every weekend, while running around the store in between customers, cleaning this and stocking that. Sometimes, during the entire eight hours, the forecourt was never empty of cars.

And inside, the lines of waiting customers never cease, with more marching in toward you like Attila's countless barbarian hordes over the Asian plains, and as you sip your (secretly vodka'd) slurpee, you feel like the late Roman Empire – isolated, vulnerable and afraid. You begin to realise, much like the guards that cowered behind the gates of Rome in AD 451, that every one of these brutes must be dealt with. Your paperwork, your food and your rising desperation for a date with the toilet are forgotten as these nameless invaders rape, pillage and befoul your store. You stand there horribly outnumbered, trying desperately to repel them.

After what seems like hours, you see a small break in cars and, for a blissful second, the store is empty. You rush to grab a stale sausage roll, but are instantly drawn back to the console by the harsh beep of the pump authorisation system and more impatient souls who have marched into the store from nowhere, standing at the counter ready to be served. Forty minutes later, you're still there, and some angry bogan is disputing the price of a packet of Tic Tacs, while you're starving, ready to crap yourself, and on the phone to tech support in Bangalore, the crackling voice not understanding you have a line of nine people and a headache like the siege of Rome.

Amidst this blur, things were coming to a head. Our 2IC had transferred to Taylor's Lakes and the list of new applicants for the position was thin. While I still would have preferred to drink a pint of bin juice than apply, Ralph, like a greying senator, had already begun scheming for his fourth and final campaign. After learning of his competition, he dramatically picked up his work rate and began to write angry, Machiavellian notes in the diary. He'd direct them vaguely at those in competition for the 2IC position, trumpeting his extra workload while castigating those who had supposedly left the slack in the first place.

He adopted the air of a leader, encouraging us all on to bigger and better things. Ask not what your servo can do for you, but what you can do for your servo. Someone would go to mop the floor and Ralph would corral them, questioning their ratio of water to detergent, following them out onto the shop floor while suggesting which area to complete first, so as to achieve the least amount of customer walk-over possible.

In the end, the 2IC job went to another woman from our site who had been there all of five minutes, and, face still smarting from this final slap, Ralph finally accepted defeat, spiralling into a fierce but short-lived depression. He requested to be transferred to a nearby site, and then soon quit and got a factory job. A *fluoro person*, of all things. In a bizarrely diametric kind of poetry, it sort of made sense. Either way, the golden era of our servo had now come to an irrefutable end. The court had lost its jester.

In Ralph's absence, I moved fully to weekday shifts, thus reacquainting myself with the glorious weekend. It was 2008 and the Dogs were starting to play good footy for the first time in a decade, and I was there almost every weekend, screaming like an entitled banshee at the umpires. Then there were gigs, clubs, more music festivals, sex in beds instead of storerooms and copious amounts of drinking beer in the fading wash of afternoon sun with mates.

But as always, it continued to grate. My subconscious was mewling at me constantly, like the disembodied voice of a haunted house: *Get out. Get out. Get out.* I had a dream (well, more of a nightmare) around then that was disconcertingly emblematic. Everything was hazy. Strange chemicals hung in the air like roving clouds. I knew I was at work, or at least some hellish variant of it that existed in some sideways multiverse. The aisles were filled with products that were alive, their darkened maws snapping like hungry turtles.

I was fuzzy, dehydrated and trapped. While on their own, these things were nothing new, I noticed that, while I was wearing my work shirt, I was naked from the waist down. (Why do our pants never make it into our dreams?)

I inspected my now completely bare body in a nearby window and found I was covered in tattoos. Or rather, a single tattoo. It was a barcode, and it encircled my entire body like a long ribbony snake, coiling around my legs, belly and torso, my arms and around my neck and onto my face.

Paying particular attention to this marker were dozens of winged mini-scanners that flitted around me like angry techno-wasps. They scanned the barcode all over my body, and every time they did, the same horrible message would pop up on the screen in front of me, and also on the giant black price board outside; a giant obsidian finger of Satan that reached up into the broiling green clouds, calling all forms of creatures down for a look.

Their message simply said: 'discontinued product'. As with most of our specials, it was nearly out of date, so had to be cleared as soon as possible. Hundreds of creatures, attracted by the horrible glow of the store, had made their way inside for their piece of the heavily discounted pie, *ooh*-ing and *aah*-ing in their weird language of clicks and clacks about the ridiculously low prices. I woke up realising the product was me.

I was meditating at home more often, that flickering triangle of orange light bringing me a lasting peace amidst the chaos and tumult of work, but still, questions swam around my psyche, the ripples from the water lapping against my skull as I wondered about what was, and what could be . . .

Ralph came in every few nights to say hi, happier than a dog with two dicks as he bragged about how much more money he was earning and how much happier he was in his new job. He stood there, with this shit-eating grin on his face as I fought off the lines of customers, talking about the simple joy he gleaned from lifting 20-kilo boxes in a massive cold-room for eight solid hours. How it was so much more rewarding than servo work. Therapeutic, even. While I knew he was lying through his teeth, I had to admit, he *did* look fitter and more refreshed than I'd ever seen him before.

I suppose this is where you expect me to say 'fuck it', then quit in an expletive-laden blaze of sex, drugs and sausage rolls. But it wasn't like that. It never is, really. As the weeks rolled on, I simply became more and more jaded. It was September 2008, and even as the world heaved with the nascent rumblings of what would become the Global Financial Crisis, I rested easy, knowing that at least I had one of the safest jobs on earth. Outside of work, my life was fine, if a little stagnant. But at work, it all began to slow down again, little by little. After the evening rush of customers had been and gone, I sometimes had twenty minutes or so before midnight to myself to think things over.

With no more Ralph, and Nellie spending more time at the prison, I was now the wise elder of the store's halogen village. As I sporadically tried to impart my wisdom onto new employees, others would ask me now and again what I was actually doing with my life. I had not even a faint shadow of an idea. I still held the crazy dream of writing for a living in a secret backroom of my head, and almost had something worth sending off to an agent, but beyond that, I was in stasis. For some, working in a servo is a career, as they hope to move up through the ranks of a lowly customer service operator to 2IC then store manager and beyond. But as you know, that was never my dream.

After graduating from uni, I felt lost. Was a shitty job in marketing where I had to drive two hours a day and earn a pittance really better than working in a servo for similar money, where I could do the job blindfolded and be home in five minutes?

These were the questions breaststroking through my brain as I was chatting to Nellie one quiet spring night, just before midnight. I'd finished my shift and was attempting to count my till, but was caught up in her stories about the Balkans and her native Serbia.

'David, in Yugoslavia, customer service not only doesn't exist, it is actually frowned upon. You can walk into some of those shops and if you interrupt their conversation then they won't just give you a dirty look, they'll slap you.'

'It sounds superb,' I said, 'but isn't it called Serbia nowadays?'

'Fuck Serbia. When I left it was called Yugoslavia. So, it is still Yugoslavia. Lines on a map, David. And lots of bombs. That's all.'

I was dropping the coins into the till but was no longer keeping track of how many there were. I'd been spending my free nights glued to my PC, researching constantly about Europe, clicking through reams of stunning pictures of Hungary, Sweden, Spain, Croatia and Montenegro. Especially Montenegro. There was this place called Kotor which I kept coming back to. It was a walled city of red roofs hidden in the corner of an epic fjord, where the surrounding cliffs plummeted down into the blue sea, and from atop a massive fort, where goats roamed free among the green grass and strewn craggy white rocks, was the most awe-inspiring view imaginable.

This place was in my head now, as I kept dropping those 50-cent coins, now not even hearing them clatter. Nellie was staring at me again, with that tiny Mona Lisa hint of a smile.

'So, when are you going back there? For good this time.'

I paused and stopped pretending to count.

'I'm not doing that. I *wish* I was doing that, but . . . nah, I'm not.'

'Yes you are, David.'

'I am?'

'Yes. You need to. Look at me, David . . . You're young, and you won't be that way forever. After me, you've been here longer than anyone, and you've learnt all you can from this shitty little box. Now it's only taking from you.'

'Well, I sup—'

'Listen to me, David. I'm stuck here. I can admit it. This place is going to keep taking from me for many more years and that's just the way it is. But I'm tough, David, I can handle this place until I die and still with a smile on my face. But you . . . no. What you need is a life. A proper life. One that makes you happy memories, free and under the sun. Not like this crappy circus of fools.'

At this point a cross-armed man who had gathered up his provisions politely coughed to let us know he was waiting. Nellie spun on her heels and snapped at him.

'*You* can wait until we're ready. This is *important*.'

The guy, like so many before him over the years, quailed away and concerned himself with the promotional display closest to him as Nellie continued.

'Listen to me, David, and listen properly,' she repeated, grabbing me by the shoulders and gently shaking me to sharpen her point. 'Look around you at all this crap, this sugary overpriced shit. Look at all the crazy people that come in through the doors. Look at how they never change. They buy the same smokes, drink the same bottles of Red Bull, whinge about the same prices night after fucking night. They'll still be doing it five years from now.'

'Yeah,' I said, 'you definit—'

'No, I'm not finished yet, David. Just listen – and think. Look at all the crazy people who work here. And look at me, David. I'm the craziest of all. Crazy old Nellie, the old wog woman who just doesn't know when to shut up and doesn't know when to quit. I can't quit, David. I have a mortgage to pay off. And a fucking divorce. From a man who likes young hairless guys instead of me. I'm the real lunatic, because I'm still here. Why are *you* still here?'

This time I remained silent.

'I'd love to be sitting on a beach right now, or even my backyard would be nice. But instead, I work in a prison and a service station, where people yell at me and threaten me every day and I smile back and tell them to go to hell. But maybe I'm the one in hell. Although I don't know what I did to get here. You *don't* have to be here, David. I think you did, once upon a time, but not anymore. You've learnt your lesson. Your long lesson. Go. Go somewhere else before they swallow you up too.'

I stood there, my eyes darting over the hypercolours as I let it all sink in, trying to peer forward and see where I'd be, years from now, wondering if it was standing morosely on the very same contoured foam mat, asking the same inane questions, writing down the same rego numbers, thinking back to exactly what I was thinking now.

I thought of all the ways this place had changed me: all the chaos I'd observed and experienced because of this crazy job. All the people I'd seen crack, then bawl their eyes out in pure pain and frustration as they quit and then slinked away, defeated. I'd decided long ago to *never* let this place beat me, but maybe, in a way, it already had.

All those rude, demanding, arrogant and borderline psychotic fuckwits. All the rabid dogmen who would go ballistic at you, straining on their chains, spittle flying from their cracked lips as they conjured up every threat they could, releasing their rage in a filthy torrent as you stood there trapped, but thankful for your cage as you took it all in, a cringing masochistic sponge, night after vexing night. Another five years of that and what would I become? One of them? While the servo had set me on the journey from the shy, lazy mummy's boy to the rough facsimile of a man I'd become, I knew I didn't want whatever it had to offer me next.

My reverie was finally interrupted at this point by the quiet man waiting to be served who, having grown tired of the promotional display (and possibly just trying to end our conversation and pay for his petrol without incurring any further wrath from the crazy Serbian woman behind the console), stared at me, assuring himself he had my attention, and then said, 'Sounds to me like a good time to get out, mate.'

Nellie turned and beamed at him. 'Now you've done something useful. What was your pump number again?'

Later, as I walked toward the door to go home, Nellie's hermetic voice followed me out into the night.

'Keep saving, David.'

I stopped in shock.

'How did you know I was saving? And besides, I haven't even decided if I'm going yet!'

'Yes you have. Goodnight, David.'

CHAPTER 24

INTO THE LIGHT

For the next few months, I was a toiling halogen ghost. As 2008 slipped into 2009, I averaged six and a half days a week, covering every shift available all over Melbourne's west, pressing buttons, smiling and shovelling dosh into my bank account like coal into a hungry locomotive. In fact, the only person I met in that time was an important one.

She was so much like me it unnerved me. After a few all-night conversations tallying up our many commonalities, we naturally christened each other 'twin'. She was a writer halfway through her first manuscript, who'd just graduated from uni. She also despised her Sisyphean retail job, replete with its own fluorescent lights, rudeness and chortling entitlement, but more than that, she took the piss out of me with relentless elan and could finish my sentences with startlingly repetitive ease.

Like me, she was a little lost, frayed around multiple edges, but all the same enthralled by those ephemeral moments that most people simply ignored. Like stepping onto the early-morning dewy grass with bare feet, or sitting on a wild deserted beach in a national park with a beer, at dusk, watching the roaring waves turn from blue to black. Finding someone who inhabits so many colours and shapes of your weird little world is the best Venn diagram there is.

She was kooky, sardonic and incisive. And she was moving to London and asked me to follow her. Finally, I had a proper reason to leave the circus. For good. Due to Dad being born in Manchester, I got a British passport, then immediately booked my flight, spending a solid hour sat dumbly in my computer chair staring at my monitor, shocked and giddy with the heft and actuality of my approaching adventure.

Generous in my newfound purpose, I gave work two-months' notice of my retirement. Barb took the news fairly well. Sort of. After I told her my plans, she deflated slightly, lost for words for the first time I could remember. Knowing I'd threatened it so many times before, she asked me, several times, if I was actually serious about it. Seeing I was, and noting the excitement and pure happiness on a face that had for so long only smirked and squinted, she softened suddenly, ignoring the beeping register and irate customers, as both relief and sadness rose to her own face.

She put a friendly paw on my shoulder and smiled crookedly, saying, 'Well, Trouble . . . I suppose it's high time you go and terrorise somewhere else,' before wrapping me in a bear hug and squeezing fiercely until there was no air left in me at all.

For those last nine weeks I paced slowly up and down the aisles as I tried to catalogue everything, to summon up some record and reverence for the last six years of my life. But I soon saw that nothing of significance remained on those steel shelves. It was the same old sugar and caffeine, just tweaked for efficiency and wrapped in shiny new sophistry from distant marketing departments. The real memories were in my head, and now, finally, all down on paper, which I'd printed out and gathered up the cojones to send off to a literary agency, come what may.

Soon enough, I was into my final week. Having said my goodbyes to everyone I worked with, I asked Barb for my last ever shift to be a graveyard, so I could properly farewell all the tottering malfeasants and drugged-up ghouls that had made every damn night Halloween and kept me company over the long desolate stretch of addled years.

I walked in on the stroke of midnight to find Nellie, smiling at me as she finished serving a beady-eyed taxi driver.

'Your till's in, David. Are you ready?'

'Fuck yeah.'

'To serve or to leave?'

'Both, you grinning fool.'

'I'm smiling for you, David.'

'I know,' I said, 'I'm smiling too.'

'I think you'll be doing that all night. I have to go now, David, so I'm going to hug you now,' she said as she grabbed me and squeezed fiercely, before turning away to the door, saying, 'I want a postcard from Yugoslavia,' over her shoulder as she disappeared and left me with my final night.

I walked around aimlessly, wondering what I should do. Steal almost everything in the store? Trash the place? Set the price of petrol to .01 cents per litre and ring all my mates?

The store was eerily quiet, so much so that I looked at the half-empty shelves and wondered aloud to myself what the current graveyarders did all night. I went and grabbed a beer from the fridge, from a six-pack of Stella I'd brought with me to celebrate my impending freedom, walking the aisles and staring out through the glass, almost willing people to show up for what may have been the only time ever.

For ages, no one came. Only a shy emo-goth kid wanting a Kinder Surprise and a new and even less talkative milkman, who only grunted at me after depositing my milk into the fridge. As I was turning to go back to the office and my meditation, the doorbell rang so I opened the door for the craggy figure of Randall. He approached the register and I prayed that he was lucid, if only to be able to say goodbye properly to the uncontested avatar of all Gumbletania.

But as so often in the past, he was gone. I saw it in his milky eyes as he walked half toward me and half toward the pie warmer.

'The spirits are singing. The angels are singing. The wizards are casting their spells into the night.'

'What spells are they casting, Randall? Why are the angels singing?'

'They're glowing. *You're* glowing. Like the sun rising in the sky.'

I smiled as I sensed some consciousness breaking through. 'Randall, I'm leaving tonight. For good.'

'The sky, the sky is full of good spells today.'

And it was. Eventually, customers showed up. They were sallow and leaden, and they shunted in against the charcoal sky, their headlights glazing the pumps in a sour gold wash, tired soldiers of everything I'd finally unshackled myself from. Their morose parade of whinging couldn't dilute my rising, somewhat tipsy, joy – not now, as the first light of day pearled salmon and euphoric through the branches of distant gums, and somewhere a kookaburra was cackling jubilantly, as if to herald my long-awaited liberty. I served each customer in a daze, gulping down the last of my beer right at the register, my gaze never properly leaving the glowing green numbers on the screen flashing rapidly toward ultimate freedom. As I was counting my till with unsteady hands, an old car with P-plates pulled in and a young guy emerged and walked into the store with a nervous look on his face.

I opened the console door for him and asked: 'It's one of your first shifts, yeah?'

'Yeah, it's my third after training. How did you know?'

'You all look the same, full of hope and eager to please.'

'Oh, okay.'

'Don't worry mate, you'll be absolutely fine.'

I stopped talking then, as it hit me that this was exactly the same situation I'd experienced more than six years ago on that foggy night full of fear, incompetence and madness. The only difference now was that I was Siva, the proud lion standing over his cub and showing him that there was nothing to fear. Not anymore.

After I got him set up on his till, he'd been serving for a few minutes when a wild-eyed truckie came barging in, barking, 'Why isn't the diesel pump fucken working!?'

It shook him, my cub, as he gazed into the jungle and saw only glowing eyes and sharp teeth. After I politely explained to the idiot how to get his petrol, my cub turned to me and asked worriedly if 'all the customers were like that one'.

I was just about to answer him when one of my old regulars burst in through the parting glass in a syncopated chorus of beats and rising synths pouring from his iPhone. He was still awake and off his dial on pills, painted by the morning sun, dancing manically to the fridge to grab two 1.5-litre bottles of Mount Franklin and then stopping by the console to load up on chewy.

Eyes like black dinner plates, he placed his phone on the counter, shut his eyes and surfed his hands through the air, savouring the trance song's satin vocals and rising strings, as he waited for the beat to drop, before finally pausing the track. He shook my cub's hand and beamed at him. 'How are ya, mate?!' he asked, as he looked down at the chocolate bars in mock distress, then straight back up with a cunning grin that made my heart smile, as I knew what was coming.

'So many choices, hey?' he said to my cub. 'I'll be here all damn morning! What's your favourite, mate?'

My cub said 'Cherry Ripe', of course, after which my regular grabbed three of them and dumped them on the counter along with a $20 note and an even bigger grin.

'The Cherry Ripes are for you, mate, enjoy! And you can keep the change, too! Have yourself a top fucken day!' As he winked at me and danced out of the store, my cub turned to me in a mixture of wonder and confusion and I leaned back against the smokes cabinet and smiled again.

I watched him serve customers for a few more minutes as he munched happily on the first Cherry Ripe, growing rapidly in confidence. I sighed happily in the knowledge that this crazy cycle would continue long after I was gone. I spent the next few minutes showing him briefly around the store, smiling at his wide-eyed expression as he rode the sugar high and eagerly soaked in everything I'd grown to despise with brand-new eyes.

I said goodbye then. For the last time. I gave my young cub a thumbs-up and walked out slowly from my cage with wet eyes. Past the anti-jump protection wire, past the chewing gum, the stale sausage rolls, the churning slurpees, the glittering dragon's horde of chocolate bars, the wonkily stacked *Herald Suns*, the leering porn, the toxic roast chicken rolls, the buzzing energy drinks, the oils and lubricants in their rainbow rows of colours, and all the invisible vestiges of madness, rage and utterly beguiling strangeness that had accumulated in my years of brightly lit servitude.

And then outside, as those glass auto-doors juddered open one last time. I walked out into the light – the *real* fucking light – and stopped on the edge of the oil-stained forecourt and stared straight up into the morning sunrise, the nascent light striping my work shirt with its orange spears, and filling my heart with a fierce unyielding joy as I walked on toward a new life. One I prayed was half as fine as the one I'd just lived.

EPILOGUE

Opposition to the servo is fading. Milk bars have vanished into a sepia-toned past. The few that remain are sad structures, cannibalised houses that are part mini-mart, part fish 'n' chip shop, manned by an elderly woman watching her customers slowly dwindle. Meanwhile, the servo powers on, strengthened by more cars on the road than ever before and tired citizens acquiescing to discount dockets and brightly crafted convenience. The Big Four fuel retailers own or control 75 per cent of fuel sites in this country. Years ago, petrol was under 95 cents a litre. These days it can reach $2.20 without breaking an oily sweat.

While the rapacious little garden gnome who ran our country for a decade was busy shampooing his eyebrows, the big servo companies were busy discovering new ways to rob us all blind. And as the behemoths of India and China snap their chains and cry out in ravenous need, oil prices only have one direction to go.

Or maybe not. Beyond his side job of posting dad-joke memes on Twitter/X and sending giant cocks into space, Mr Musk may just prove to be the forerunner for Netflix to the servo industry's Blockbuster. Electricity is, after all, outrageously cheap, and if everyone can charge their EVs in their garage every night for one-fifteenth of the price, then servos are starting to look as anachronistic as MySpace.

What this means is anyone's guess. Will the servo die along with oil or will it mutate into some new palace of vital convenience? I don't pretend to know the answer. What I do know, though, is that the servo is a garish modern shrine to both our laziness and our shallow-minded search for quick comfortable solutions.

As drive-offs increase, and technology marches inexorably on, machines will prevail. While many of the people who work in servos have already been beaten into robots by Orwellian strictures of control, half the store is already one big vending machine. Eventually, the workers and the range of products will dwindle, replaced by a 'super product': some *Soylent Green*–style amalgamation of sugar, caffeine, SSRIs, nicotine, chocolate, paracetamol, amino acids, salt, fizz and vitamins.

They'll install a massive AI dispenser that will take up an entire wall, displaying the latest happenings from the Metaverse while mashing all these things together into a pulpy low-GI, 99-per-cent-fat-free ball, scanning your retinas or inserted chip, instantly charging your UBI-fuelled CBDC account, then depositing it fully wrapped into the waiting slot to the sound of your favourite music so you can get your modern soma.

Interaction between humans will be phased out. Your addictions taken care of, your self-driving EV will then soundlessly leave the pit stop, with its spectral holographic advertising boards painting new purchasing suggestions onto the sky, as a distant pang of melancholy projects a faded image somewhere in the dusty catacombs of your hippocampus of a comatose Indian student blinking tiredly at you as he morosely pesters you to buy two packets of Extra for three bucks.

What I'm trying to say is that servo people could be on the endangered species list before you know it. That friendly guy who used to call you a taxi, who used to let you off 25 cents, who kept an impressively straight face after seeing the menagerie of questionable sexual partners you dragged home from your local nightclub, may soon be no more.

So next time you get a 3 am sugar craving, or maybe just when you're off your face on powerful narcotics and desire an audience for your id's upswelling manifesto, come on down to the bright box of colours and say hello. Stay and chat for a while. Or even just stand there and watch.

This circus is free. And it's here every night of the year.

ACKNOWLEDGEMENTS

My sincere and heartfelt thanks to every human who helped bring this crazy little goblin of a book into the world.

Thank you, first and foremost, to Sophie Hamley, not only for your incredibly helpful notes but also your unwavering belief in and championing of the many incarnations of this strange evolving beast over the years. We got there in the end!

Thank you to everyone at Hachette, especially Emma Rafferty and Susin Chow, both for reining in my addled flights of fancy and helping the book shine clearer than a freshly squeegied windscreen.

Thank you to the talented Josh Durham for such an incredible cover that perfectly summarises *Servo*'s 'mad dark circus' vibe.

Thank you Barb, for not firing me, even though I gave you damn good reason to, several times a week.

Thank you to my long-suffering servo comrades, some of you still toiling away in your halogen cages and smiling (with teeth). I raise my slurpee in solidarity with your struggle.

Though I never thought I'd say it: thank you customers. You crazy grinning goons. Where would I be without your enduring lunacy? You are the true and generative stars of this dark and roiling circus; I just stood there nervously with a long snake of paper and a pen.

And thank you Mum, for not only putting up with that pale, sleep-deprived monster, but unfailingly leaving him glad-wrapped dinners in the fridge. Even fried in radiation, every one of those meals tasted completely like love. ♥

David Goodwin survived weekend graveyards in servos
for six interminable years – way too long to stay anything
approaching sane. He is, thankfully, no longer a daysleeper
with a halogen tan, but still maintains a ruinous predilection for slurpees, chocolate Big Ms and sausage rolls with
too much tomato sauce. He is a published poet, holds a
dual Advanced Diploma in Advertising and Marketing
and, these days, revels in having a somewhat normal
circadian rhythm.